How Dogs Love Us

A Neuroscientist and His Adopted Dog
Decode the Canine Brain

GREGORY BERNS

Little
a

Published by Little A, New York

www.apub.com

Author photograph © Bryan Meltz

ISBN 9781477800874

Printed in the United States of America

For Lyra

Contents

When the Man woke up he said, 'What is Wild Dog doing here?' And the Woman said, 'His name is not Wild Dog any more, but the First Friend, because he will be our friend for always and always and always. Take him with you when you go hunting.'

— RUDYARD KIPLING, *JUST SO STORIES*

Dress Rehearsal

EMORY UNIVERSITY, ATLANTA, GEORGIA
JANUARY 2012

CALLIE WAS DANCING in the lab. Zooming from person to person, the little black village dog with the energy of a rocket knew that all the months of training had led to this moment. Her eyes sparkled with life, and her rat-tail wagged side to side with such intensity that her head moved in exactly the opposite direction. She was ready.

Let's get on with it!

Callie's excitement was infectious. Everyone in the lab wanted to see the experiment we were about to perform, mostly because nobody thought it would work. Could we really scan a dog's brain to figure out what it was thinking? Would we find proof that dogs love us?

With the team assembled and ten minutes until scan time, we headed to the hospital. Dogs, of course, were not allowed on campus, and here was a very special dog marching across the quad with a dozen people in her entourage. I carried the backpack full of treats and supplies, Andrew toted the computer that would record the timing of the experiment, and Mark hauled the plastic stairs that would let Callie walk up into the MRI — she would have to do that on her own. Everyone else tagged along, snapping photos and texting their friends: the Dog Project was actually going to happen.

Students, stuck in their lectures, stared out classroom windows as Callie led us all to meet her destiny with a big-ass magnet.

We entered the MRI room through a secret entrance to the hospital.

Even though the Dog Project had already taken on a circuslike atmosphere, there was no need to alarm the patients by parading Callie through the hospital corridors. I pulled shut the massive door, which was clad in copper to keep out stray electrical signals. It made a tight seal, almost like an air lock. With the room secure, I let Callie off-leash.

With nose to the ground and tail held high, she trotted around the MRI scanner, making several circuits. Curiosity satisfied, she exited the magnet room and checked out the control room. Despite the hospital setting, the floor was filthy. Several years earlier, a janitor had attempted to clean the MRI room. Imagine his surprise when the floor buffer levitated off the ground and crashed into the bore of the magnet. Ever since, the janitorial staff had been forbidden from entering the facility. The cleanliness had subsequently declined.

Callie, of course, found every crumb of organic matter that had, at one time, been edible.

Before we could do any brain scanning, Callie would have to go into the magnet. Normally, magnetic fields are imperceptible to us. But the MRI creates a field sixty thousand times stronger than Earth's magnetic field. You definitely feel it. As you approach the center of the MRI, the magnetic field increases rapidly in intensity. If you moved a piece of metal through the field, an electrical current would be induced. The same thing happens when a person moves through the magnetic field. The field induces small electrical currents in your body. These currents are most prominent in the inner ear, creating a slight spinning sensation as you are moved into the center of the magnet. For some people, though, it can create a nauseating sense of vertigo.

Up until that moment, the thought had not occurred to me that dogs might be more sensitive to the magnetic field than humans. We were about to find out.

I placed the portable steps at the base of the patient table. Callie sniffed them but showed no interest in climbing them. She continued to trot around the room, curious about every nook and cranny. Time to bring out the hot dogs.

That got her attention. Unable to resist the scent of hot dogs, she padded up to the top of the steps, but once there, she balked at climbing onto the patient table. Of course, I could have picked her up and put her there, but it was important to remain faithful to our ethical principle of self-determination. Callie had to do it of her own free will.

The MR techs started laughing. How could we do an MRI if the subject wouldn't even get on the table? But I knew Callie would eventually come around. The environment was new and exciting. Once she settled down, she would focus on what she had already learned.

After five minutes of walking up the steps and jumping off, Callie tentatively placed her paw on the patient table. With great enthusiasm I encouraged her to keep going.

"That's it, Callie! Good girl! Want more hot dogs?"

She got it. Once up on the table, she saw it wasn't scary at all and that there was a ready supply of hot dogs. Now she had to go into the bore of the MRI.

I had already secured the foam chin bar in the head coil, located dead center in the magnet tube. Now I placed, Hansel and Gretel style, a trail of hot dogs leading from the entrance of the MRI to the head coil. Without a thought, Callie continued walking down the patient table into the bore, lapping up hot dogs as she went.

My colleague Lisa, who was filming the event, gasped in excitement at the sight of a dog walking into the MRI.

I quickly circled the scanner so I could face Callie from the other end of the bore. She was crouched down in a sphinx position just short of the head coil. Her tail was swishing back and forth. I reached in with a hot dog in my hand and immediately felt the room spin.

Callie saw the hot dog and scooted forward into the head coil.

"Good girl!" I said with my highest, most excited voice.

She took the hot dog and backed up a little bit, but she didn't leave the bore. With a steady stream of hot dogs, Callie quickly adapted to the new environment and was soon happily eating treats while nestled in the head coil. There was no indication that the magnetic field bothered her.

With Callie comfortable in the magnet, we had accomplished the first goal of the session. Since that had been relatively easy, it was time to see how she would react to an actual scan.

The scanner software was created for human subjects, so it had no way of knowing that Callie was a dog. Inputting an accurate weight of the subject was the most important piece of information because that determined how much radio power the scanner would emit.

Too much power would cook her like so much meat in a microwave.

With hot dogs, I once again coaxed Callie into the MRI. When she was comfortably settled in the head coil, I gave a thumbs-up. The scanner made a series of clicks and hums as it revved up.

Callie's eyes narrowed.

Then, like the onslaught of a thousand bees, the scanner started buzzing. This was the initial preparation phase, called *shimming*. The scanner automatically adjusts the magnetic field to compensate for the distortion caused by whatever is placed inside. Normally, shimming takes a few seconds, but with Callie inside, the buzzing continued. Even though she was wearing earmuffs, she wanted no part of it and headed for the exit.

I waved my arms back and forth, signaling the MR tech to abort the scan.

"What was that noise?" I asked.

"Shimming," he said.

"Why was it going on for so long?"

"The scanner was having trouble compensating," he explained. "Probably because it expects a human."

We hadn't thought of this. We hadn't even recorded the shimming noises for our training sessions. We had assumed they would take only a few seconds, a minor blip compared to the lengthy functional scans that would follow. Callie reacted to these novel sounds as any dog would: she got scared.

We tried a dozen times, but Callie scooted out as soon as the scanner started buzzing. We even tried starting the scan before she went in, figuring that if she got used to the ambient noise, I could coax her into the head coil. Eventually, with enough repetitions, the scanner was able to cobble together a crude compensation for her canine form.

Next up were the functional scans. These are a series of scans that individually take about two seconds to capture the brain. By continuously acquiring these functional scans while Callie was in the MRI, we could measure changes in her brain activity and figure out what she was thinking. At least, that was the plan. Finally, we would do a structural scan, which is a high-resolution picture of the brain used to identify brain anatomy.

It was tough for her. The earmuffs kept sliding back, exposing her ears to the full onslaught of the noise. Even so, Callie managed to hold her head in position for a few seconds at a time. We stopped the scanner after three minutes' worth of scanning. That, we felt, would be enough to evaluate the quality of the data.

Before she got too tired, we decided to make one attempt at a structural image. The structural scan takes thirty seconds, and Callie would have to hold still the entire time. After the scan, she bounded out of the magnet and pawed off her earmuffs. She jumped up and licked my face and then ran over to Lisa, who gave Callie a big hug.

"What a good girl!" she exclaimed.

We all went into the control room to see what the images looked like.

The structural image looked remarkably good. There were ghost images throughout, which occur when the subject moves, but it was clearly recognizable as a dog's brain. The functional images were a different story. Out of 120 images, only one contained anything that looked like a brain. Mostly they were jumbles of digital snow with an occasional eyeball peeking into the field of view.

I hugged Callie and said, "I'm so proud of you." But in reality, I didn't know if this was going to work.

The next scan — with Callie, the other dog, McKenzie, and the whole entourage — was in three weeks. I hoped we could figure it out before then. If we didn't, I would have to pull the plug on the Dog Project and acknowledge that the naysayers had been right: you can't scan the brain of an awake dog.

How Dogs Love Us

1

Dia de los Muertos

TWO YEARS EARLIER

EVERY NOVEMBER 1, I push aside the remains of the Halloween candy and erect a shrine on the dining room table.

I begin with a vase that Kat and I bought in Mexico on our honeymoon. It's a cheap thing, with a stylized owl painted on one side, but the vase has somehow survived multiple moves across the country, and I have come to value it for its resiliency rather than its beauty. It also provides the necessary ethnic authenticity for the ritual and functions as an ideal centerpiece to prop up photographs.

We keep the photos in a drawer all year long, only to be brought out on this day. Kat and I surround the vase with them: pictures of family members who have passed away over the years. Then, to complete the offering for their spirits, we scatter a cornucopia of the sweetest, most delicious baked goods.

Our two daughters, Helen and Maddy, had never questioned why we did this. They had, after all, lived with the ritual all their lives. But when they achieved the age of enlightenment, preteen-hood, they

realized that celebrating Dia de los Muertos — the Day of the Dead —
was not a normal thing to do. At least not the way we did it.

We included the dogs.

Although I had grown up with dogs, it wasn't until I finished medical
school that I had the opportunity to acquire a dog that I could truly
call my own.

Kat and I had been married for five years, and we were putting off
children until I completed my training. So, in celebration of complet-
ing my first year of medical internship — a grueling year of hundred-
hour weeks — we answered an ad for puppies. Pug puppies, actually.
I note this with some qualification, because to many, pugs are a gro-
tesque distortion of the canine form. Of course, Kat and I didn't see
them that way. Their large heads, with pushed-in noses and bulbous
eyes, were almost human — a sort of baby substitute.

We named our new puppy Newton.

Like all pugs, Newton's face was brachycephalic, meaning short-
nosed, but his was foreshortened in the extreme, with his nostrils
forming mere slits. He was what breeders call an apple head because
of the taper of his skull. His panoply of malformations only made
him more endearing to us, and his constant snuffling and snoring
became a welcome background noise to our lives. At night, he slept
with his unusual dome nestled in my armpit.

Newton was smart and energetic — and a prankster. He would
chew the tags off our clothing only to vomit them up an hour later.
Once, he got into a bag of chocolate-covered espresso beans, prompt-
ing a panicked call to poison control. I could hear laughter before
they assured the newbie dog owners that their precious pug would be
okay.

We immersed ourselves in pug culture. We socialized with other
owners, all of whom echoed the analogy between pugs and Lay's po-
tato chips: you can't have just one. So it was no surprise that within

a year we had adopted two more pugs. Six-year-old Simon was the opposite of Newton: simple, sweet, and dimwitted. Dexter tipped the scales at thirty-three pounds after a lifetime of being fed hamburgers by a truck driver who took him everywhere but could no longer care for him. He was like Jabba the Hutt, waddling his rolls of skin around the house. Mostly he just liked having his chin rubbed.

Newton.
(Gregory Berns)

Dexter was the first to go. Helen was three years old and Maddy had just turned two. That was the year we started celebrating Dia de los Muertos. At first, Dexter was the only dog spirit, and Kat and I began the tradition of leaving dog treats for him. Simon followed the next year.

As much as we loved Newton, the house just didn't seem right with only one dog. It didn't take long for the girls, especially Helen, to ask for a dog of their own. But they wanted a big, fluffy dog that they could play with. (Pugs in their later years don't play very much.)

Not long after Simon's passing, a respected breeder in our neighborhood had a litter of golden retriever puppies become available. Only three were left when we visited. We took home the only female, a sweet bundle of light golden fur. We named her Lyra, after the protagonist of Philip Pullman's wonderful book *The Golden Compass*.

Lyra settled into the house easily. She epitomized the affability that has made golden retrievers such a popular breed. She never protested when the girls' friends climbed on top of her, and she got along with every dog in the neighborhood, even a pair of irascible Jack Russell terriers that lived down the street. In part because of her easygoing, submissive personality, and in part because of her flowing golden mane, Lyra became a popular fixture in the neighborhood. Kids

Helen, Maddy, and Lyra.
(Gregory Berns)

would run up to her to embrace the walking teddy bear. And Lyra would just grin.

With time, the jet-black muzzle of Newton's youth faded completely to gray, only his ears retaining some dark pigment. Most of his teeth were rotted out from a lifetime of mouth breathing, and his fountain of energy dwindled to a trickle. By the time he was fifteen, he suffered from a slowly progressive deterioration of the spinal cord. Newton eventually lost the use of his hind legs, requiring a doggy wheelchair. He soon lost control of his bladder and bowels too. Never in his life had Newton had an accident in the house, and his look of shame, as he struggled to crawl away from his mess, told us that it was time.

As I laid Newton in his grave, he gave one last snort. I knew it was the remaining air in his lungs being expelled, but I still like to think it was his soul crossing the Rainbow Bridge into the mythical land where pets and humans are reunited.

Even though I didn't know it at the time, the seed for the Dog Project was planted with Newton. It was Newton's spirit that continued to hold the greatest power over me. We had shared fifteen years together, and I had never really known what he was thinking.

What I really would have liked to know was whether he truly returned my feelings toward him. But I would have needed some sort of canine brain decoder to know whether he loved me.

A few months after Newton's death, the kids were on spring break. Kat and the girls decided to take a trip to the animal shelter.

The first hint that something was afoot was a text message from Kat. She attached a blurry photo of a dog slung over her shoulder. It was a long, skinny thing with sticks for legs. It was so black I couldn't make out any details except for its four white paws. Its head looked like an anvil with one ear pointing straight up and the other flopping over its face.

Game over. Once they had stepped into the animal shelter, there was no way Kat and the girls were coming home without a new dog.

The girls quickly set to drawing up a list of potential names, but for the first day we called the dog by her shelter name: Little Miss Piggy. The shelter had a new theme every week for naming the animals that came into its care, and this happened to be Muppets week. Given the randomness of the naming protocol, you wouldn't think there was any significance to the shelter name. But Little Miss Piggy would soon prove otherwise.

The kids weren't fans of the Muppets, so there was no question that the shelter name had to go. Plus, it was just too long. I couldn't imagine standing on the porch yelling, "Little Miss Piggy, come. Little Miss Piggy, come!"

Our new dog was a mystery. We had no idea where she came from or how she had ended up in the shelter. While she wasn't afraid of humans, she did seem to prefer curling up with Lyra. Maybe she hadn't had much human contact. The shelter estimated her age at nine months. Most dog experts recommend socializing puppies to humans by six weeks. So although it was a bit late for our new pet, the lack of fear meant that even if she had had little human contact previously, at least she hadn't been abused.

In the end, it was Maddy who came up with the name.

"How about Calypso?" she asked. Maddy was devouring everything she could find about Greek mythology at the time. Calypso was a minor goddess in the *Odyssey* who prevented Odysseus from leaving her island in order to make him her husband. This went on for seven years before Athena intervened and returned Odysseus to his true love, Penelope. In Greek, Calypso means "to cover" or "to hide," and given her black coat, this seemed appropriate. So Little Miss Piggy became Calypso.

Callie for short.

Callie on high alert.
(Gregory Berns)

Weighing barely nineteen pounds, Callie was about twelve inches tall and eighteen inches from nose to rump. Like all mutts, her tail curled up in a C over her back. Her ribs were clearly visible.

Helen went online to figure out exactly what she was.

"Jack Russell terrier," she said pointing to an image on the computer.

"The color isn't right, and she's taller than that," Maddy said.

"She's obviously a terrier of some type," Kat pointed out.

I thought Callie looked a bit like the old RCA dog that peered into the cone of an early Edison phonograph. The lineage of that dog, named Nipper for his fondness of biting, is unclear. Some say he was a Jack Russell, others say a fox or rat terrier. Callie almost resembled a Manchester terrier. But Manchesters are always black and tan in color, and Callie was black and white, which ruled against a pure Manchester pedigree. Some say that the Manchester, or black-and-tan terrier, as they were once called, was crossed with a whippet in the nineteenth century. Whippets are like little greyhounds, and this crossing surely increased the terrier's speed. As we soon found out, speed was one of Callie's defining traits.

In fact, Callie was the fastest dog I had ever seen.

The first time we let her in the backyard, she established a perimeter by running circuits just inside the fence. Most of the yard, though, was thick with English ivy, which in the lush southern climate grew in knee-high thickets. Callie would run full tilt, alternately leaping over the ivy and then diving beneath its heavy leaves. As soon as she got to the fence line she would tunnel under the ivy, tracking the edge of the fence. Like a torpedo, all you could see was a bulge moving through the ivy at high speed, only to explode out in a leap of joy. With her back muscles flexing and unflexing, she ran like a cheetah.

On top of her speed, Callie exhibited a laserlike focus on squirrels. Once caught in her field of view, a squirrel would be chased up a tree in sheer terror with Callie scrabbling at the bark. The treeing of squirrels wasn't the main characteristic of terriers, and although in all likelihood Callie was a mongrel, I couldn't help but hold on to the fantasy that we had found an unrecognized purebred at the shelter.

The treeing of squirrels turned out to be the key to decoding Callie's heritage. Not being native to Georgia, I was unaware of a breed peculiar to the South: the treeing feist. According to the American Treeing Feist Association, the treeing feist, or mountain feist, existed in the southern Appalachians long before rat terriers were brought to America. While terriers were bred to catch vermin, feists were bred to hunt. And while squirrels are their primary prey, the feist will gladly hunt raccoons, rabbits, or birds. With longer legs than terriers, feists are built for silent speed. They live to tree a squirrel until its owner comes to catch it. The feist has a storied history intertwined with the beginnings of the country. George Washington wrote about them in his diary, and Abraham Lincoln even referred to them in a poem.

Over the next two days, Callie's personality began to emerge. And even though we had changed her name, somehow the shelter had been closer to the mark. She loved to eat.

Every day when I came home from work, Callie would burst into the room and jump up and down like a pogo stick, wagging her tail furiously, her eyes filled with pure joy. But on the fourth day, when I came home, she just lay on the rug, hardly moving at all.

"What's wrong with Callie?" I yelled out to Kat, who was busy helping Helen with homework.

"What do you mean?" Kat asked.

"She's just lying here on the floor." This brought everyone running into the room.

Maddy covered her mouth as her eyes began to tear up.

"What's wrong with her?" Helen asked.

Callie just rolled on her side and began to whimper.

I kneeled down to see what was wrong. Her belly was bloated out of all proportion to her rail-thin figure. When I touched her tummy, she squirmed away and made a little whine.

We immediately set to looking for what she might have eaten. I expected to find the tattered remnants of a shoe or one of the kids' toys. Ten minutes of searching yielded nothing, and Callie just seemed to be getting worse. Moving from sitting to standing to lying down, she was unable to find a position that didn't cause her pain.

Finally Kat yelled out from the kitchen pantry. "I found it!"

This is where we kept the dog food. Over the years, we had learned that it was cheapest to buy fifty-pound bags. For ease of access, we stored the food in a large plastic bin on the floor of the pantry. With the lid on, the bin had successfully kept the dogs from helping themselves. Until now.

The lid was pried off, and a few bits of kibble were scattered about. Callie had somehow figured out how to open the container and had gorged herself into oblivion. There was still plenty of food left in the bin, but then again, none of us knew how much had been there before she started. The bits on the floor indicated that she had eaten so much that she hadn't bothered to pick up the remaining scraps.

Helen was on the verge of panic. "We have to take her to the vet!"

I looked at the clock. Past six. Kat was thinking the same thing: after-hours emergency visit. This was going to be expensive.

Callie was such a thin, little dog, and her belly was like a balloon. It was hard to imagine how all that stuff was going to make it through. Maybe she ate so much that she tore her stomach. Was that possible? I had heard of such things happening in humans, but never in dogs.

"Do you know what happened to the rawhides?" Kat asked.

"What rawhides?"

"The pack of rawhides I bought yesterday."

As Callie writhed on the floor, we both knew the answer.

We headed to the emergency veterinary clinic. This was a fully staffed, multispecialty hospital, manned 24/7 and arrayed with the latest medical technologies. But unlike a human hospital, this was strictly pay in advance. Two hundred dollars to open a tab.

We weren't exactly sure what else Callie had eaten, so the first order of business was an X-ray.

"You see that?" asked the vet as she pointed to what looked like the silhouette of a dog filled with popcorn. "That's all food. The good news is that there aren't any foreign objects."

"And the bad news?"

"She can't really drink anything in this state. If she gets dehydrated, the food could turn to concrete in her stomach, which will make it very hard to pass. I recommend we give her an IV to keep her hydrated and keep her overnight."

None of us wanted to leave our new pet overnight in the hospital. Helen summed it up: "Daddy, she just came from the shelter. She's really scared."

"Can't you just make her vomit?" I asked the vet. That was clearly not where she had intended the conversation to go.

"We can try," she replied with some resignation. "But it doesn't usually work at this stage."

Since there wasn't much of a risk from trying, Callie received an injection of apomorphine, a potent emetic. Within five minutes she began retching. But, as the vet had predicted, nothing came up. Callie just looked confused and frightened.

There wasn't any choice. With tears all around, we said good-bye to Little Miss Piggy and trundled out of the hospital. Even though she had been with us only a few days, I couldn't help but feel that we had somehow failed her. What kind of pet owners were we if our new adoptee landed in the hospital within the first week?

The next morning, the hospital called to say that Callie's vital signs were stable but that she hadn't passed anything yet. They recommended keeping her another twenty-four hours.

"Can we take her home?" I asked.

"We don't recommend it."

Helen pulled at my sleeve, begging me to pick her up.

Kat and I figured that if anything bad was going to happen, it would have happened already. Besides, Callie had the benefit of being rehydrated by IV, which we hoped would keep her tanked up until the food made its way through.

At the hospital, we had to sign Callie out "against medical advice." Yes, we were very bad dog owners. When we got home, Callie bounded into the house as if nothing had happened. She drank a bunch of water and ran outside to porpoise through the ivy.

We had medical insurance from the animal shelter through the first thirty days of the adoption, but the insurance company denied

the claim. Some fine print about covering only foreign-body inges-
tions, not pathologic overeating.

It didn't matter. I was just grateful Callie was okay. And she would
soon change my life, helping answer my questions about what Newton had felt and eventually revealing clues to the deeper question:
What are dogs really thinking?

2

What It's Like to Be a Dog

THE IDEA OF SCANNING DOGS' BRAINS didn't occur to me all at once. As with most scientific developments, it started as a series of random thoughts and inferences that eventually led to an aha moment. While Newton's death planted the seed of an idea, it was my own discomfort around groups of people that helped it grow.

For the past fifteen years, my lab has used brain-scanning technology to understand how the human reward system works. The main tool that we use is magnetic resonance imaging, or MRI. About the size of a car, an MRI scanner is pretty much a large tube wrapped in miles of wire. When electricity is sent through the wire, it creates a powerful magnetic field that can be used to see inside of a person's brain. A standard MRI, like what you would get if you went to a hospital, takes a picture of your brain. Scientists soon discovered that if you took several pictures of the brain in rapid fire, you could see the brain in action. This is called *functional MRI*, or fMRI, and it opened the black box of the human mind. With fMRI, we can measure activity inside the brain while a person is actually doing something, like reading or doing math or even while experiencing different types of emotions. This allows scientists to figure out how the brain actually works (hence the *functional* in fMRI).

As the leader of a research lab, it is one of my duties to hold an annual lab party. You would think that this would be an enjoyable activity. Inevitably it is a source of stress in our household. The dogs don't help either.

Like me, the dogs were never properly socialized to large groups of people, something for which I take full blame. Since we don't have parties often, it seemed unreasonable to make the dogs learn how to behave in such situations. Nevertheless, one cannot completely abdicate these social necessities, as with our once-a-year gathering of lab members.

Ignoring my antipathy, Kat and the girls threw themselves into the preparations for the annual party. They brought all the chairs out of the dining room and created a semicircular seating arrangement in the family room. Nothing unusual about this, presuming that the guests are able adults who can manage conversation while eating and drinking without tables to place their food upon. It does not, however, account for dogs either underfoot, in the case of Callie, or swishing big, fluffy tails around, in the case of Lyra.

If everyone was a dog person, these parties wouldn't present a problem. In recent years, I have certainly become more selective in allowing people to work in the lab, and this includes my asking whether he or she is a dog person or, second best, a cat person. But can you really trust someone who doesn't have a pet? Despite my best efforts to fill the lab with animal lovers, I have no control over spouses and partners.

Kat wanted to lock Lyra and Callie in the bedroom when the guests arrived. The dogs weren't accustomed to being locked up, so I feigned ignorance and let them have free run of the party. As guests arrived, Callie would give a perfunctory woof. Lyra just grinned and wagged her tail excessively as the people filed in.

I could trust the dog people in the lab to keep an eye on the dogs and prevent them from swiping food, so I slipped out to help Kat in the kitchen. She was dishing up the hors d'oeuvres and pouring drinks. The team, while diverse in terms of background, was predominantly American, with the exception of one lab member from India. It was at the moment I stepped into the kitchen when he arrived with his wife.

Their entrance was marked in dramatic fashion by an ear-piercing "Eeeeeee! Eeeeeee! Eeeeeee!"

I rushed out of the kitchen. My colleague's wife, wrapped in a lovely sari, had backed herself into a corner, shrieking like a bird at the mere sight of the dogs.

This behavior baffled Callie, so she paid no further notice to her and moved on to look for food droppings. Lyra, on the other hand, found these vocalizations highly stimulating. She tracked right to the sound and starting jumping up and down and barking in what appeared to me to be a request to play. But the grimace of terror on the woman's face indicated no such desire.

I grabbed Lyra by the collar and led her to the bedroom.

"Sorry, girl. You can't play tonight."

What did Lyra think was the reason that woman was screaming? If Lyra were a person, I could have simply asked her. How else could I find out what was going through her mind?

To truly know what a dog is thinking, you would have to be a dog.

The question of what a dog is thinking is actually an old metaphysical debate, which has its origins in Descartes's famous saying *cogito ergo sum* — "I think, therefore I am." Our entire human experience exists solely inside our heads. Photons may strike our retinas, but it is only through the activity of our brains that we have the subjective experience of seeing a rainbow or the sublime beauty of a sunset over the ocean. Does a dog see those things? Of course. Do they experience them the same way? Absolutely not.

When Lyra was jumping and barking at the woman wrapped in purple, with a red dot on her forehead, Lyra experienced the same things at a primitive level that I did. Purple. Red. Screaming. Those are the sensory primitives. They originate in photons bouncing off dyes, pressure waves in the air around the woman's vocal cords. But my brain interprets those events one way and Lyra's brain another.

Observing Lyra's behavior doesn't tell us what she was thinking. From past experience, I knew that Lyra barked and jumped in response to different things. She barks when we're eating. In that context, a natural assumption would be that she wants food too. But she also barks after dropping a tennis ball at my feet. I had no comparable frame of reference for what had attracted her to the screaming woman that night at the party.

The question of what it is like to be a dog could be approached from two very different perspectives. The hard approach asks the question: What is it like for a dog to be a dog? If we could do that, then all the questions about why a dog behaves the way it does would become clear. The problem with being a dog, though, is that we would have no language to describe what we felt. The best we can do is ask the related, but substantially easier question: What would it be like for *us* to be a dog?

By imagining ourselves in the skin of another animal, we can recast questions of behavior into their human equivalent. The question of why Lyra harassed the party guest becomes: If I were Lyra, why would I bark at that woman? Framed that way, we can form all sorts of speculations for dog behavior.

Many authors have written about the dog mind, and some have

even attempted to answer the types of questions I have posed. I will not review this vast literature. I will, however, point out that much of it is based on two potentially flawed assumptions — both stemming from the paradox of getting into a dog's mind without actually being a dog.

The first flaw comes from the human tendency to anthropomorphize, or project our own thoughts and feelings onto things that aren't ourselves. We can't help it. Our brains are hardwired to project our thoughts onto other people. This is called *mentalizing*, and it is critical for human social interactions. People are able to interact with each other only because they are constantly guessing what other people are thinking. The brevity of text messages, for example, and the fact that we are able to communicate with less than 140 characters at a time work because people maintain mental models of each other. The actual linguistic content of most text exchanges is minimal. And because humans have common elements of culture, we tend to react in fairly similar ways. For example, if I watch a movie that makes me sad, I can use my own reaction to intuit that the people sitting around me are feeling the same way. I could even start a conversation with a complete stranger based on our shared experience, using my own thoughts as a starting point. But dogs are not the same as humans, and they certainly don't have a shared culture like we do. There is no avoiding the fact that when we observe dog behavior, we view it through the filter of the human mind. Unfortunately, much of dog literature says more about the human writer than the dog.

The second flaw is the reliance on wolf behavior to interpret dog behavior, termed *lupomorphism*. While it is true that dogs and wolves share a common ancestor, that does not mean that dogs are descended from wolves. This is an important distinction. The evolutionary trajectories of wolves and dogs diverged when some of the "wolf-dogs" started hanging out with proto-humans. Those that stuck around became dogs, and those that stayed away became modern wolves. Modern wolves behave differently from dogs, and they have very different social structures. Their brains are different too. Interpreting dog behavior through the lens of wolf behavior is even worse than anthropomorphizing: it's a human anthropomorphizing wolf behavior and using that flawed impression as an analogy for dog behavior.

Wolf analogies have led to many flawed training strategies based on the idea that the human must be the "pack leader," an approach most commonly associated with Cesar Millan. Unfortunately, there is no scientific basis for using the wolf's social structure as a model for the dog-human relationship.

Dogs can't talk, and we can't transport ourselves into a dog's mind to know what its subjective experience is. Where I see a happy golden retriever playfully jumping up and down, someone else might see a hungry dog planning to eat her for dinner. So what can we do to better know a dog's mind?

Although I hadn't yet made the connection at the party, I would soon realize that the solution had been right in front of me all along: brain imaging.

Because all mammalian brains have substantially similar parts, a map of canine brain activation could be referenced to its human equivalent. For instance, if we saw activation in the reward center of the dog brain, that could be interpreted through human experiments that result in similar activity. With human experiments, we have a reasonably good idea of what happened to create a particular pattern of brain activation. We know, for example, that activity in the visual part of the brain can be caused either by photons hitting the retina or by the person mentally imagining a scene with his eyes closed. Similarly, if we observed activity in the visual part of a dog's brain, and the dog wasn't looking at anything, we could reasonably assume that it was forming a mental image of something. Dogs might have imaginations too!

Mapping between the brains of different species is called a *functional homology*. It means that a subjective experience like imagination can map onto both a human brain and a dog brain. The patterns of activity in the two brains would illustrate how to transform one type of brain into the other.

Philosophers dismiss the question of what it is like to be a dog as unanswerable, but functional homologies between dog and human brains could provide the missing link. Although brain imaging wouldn't tell us what it is like for a dog to be a dog, it could provide a road map — a brain map — of what it would be like for a human to be a dog, without the bias of the human interpreter. If it worked, brain imaging could end up being a canine neural translator. We could go way beyond the question of why Lyra was being obnoxious at the party. If we could map our thoughts and feelings onto the dog brain, we could get right to the heart of the dog-human relationship: Do dogs love us?

It all comes down to reciprocity. If the dog-human relationship is predominantly one-sided, with humans projecting their thoughts onto the dog vacuously staring up at his master in the hopes of receiving a doggie treat, then the dog is not much better than a big teddy bear — a warm, soft, comforting object.

But what if the dog reciprocates in the relationship? Do dogs have some concept of humans as something more than food dispensers? Simply knowing that human feelings toward dogs are reciprocated in some way, even if only partially, changes everything. It would mean that dog-human relationships belong on the same plane as human-human relationships.

None of these questions can be answered simply by observing

dogs' behavior. They go to the heart of dogs' subjective experience of the world and, in particular, their subjective experience of us.

My colleague and his wife didn't stay long. Even with the dogs locked away we could hear Lyra barking in the bedroom above the din of the party. Nobody was surprised when they were the first to say good-bye.

Once they left, I let the dogs out. Lyra ran to the remaining guests and, in her state of excitement, puked up something foamy and green. The partiers watched in disgust as Callie darted over to slurp it up.

From the chorus of "Oooh, gross!" it was clear even the animal lovers were aghast at our dogs' behavior. An exodus ensued.

And that is why we no longer hold lab parties at our house.

3

A Fishing Expedition

WHILE THE EMBARRASSING INCIDENT of the lab party was the second catalyst of the Dog Project (Newton's death being the first), the final event that set the project in motion came out of the blue: the death of Osama bin Laden.

Every Wednesday morning, the members of my research group gather for the one sacred event of every academic laboratory: lab meeting. Regardless of the field of research, every lab in every university holds a meeting once a week, the only time when everyone, from the lowliest undergraduate to the lab director himself, has the opportunity to learn what everyone else has been doing. At lab meeting, everything is laid on the table. You hear about new discoveries, unexplainable data, and false leads.

All the research my lab does is based on MRI data. We are a "dry" lab because we don't work with chemicals or do biological experiments that require expensive containment equipment. Those types of labs are "wet" because they have specialized plumbing and air vents to prevent the release of toxic fumes or, worse, infectious microbes. Our lab doesn't even have a sink. It is simply a large room with computer terminals located around the perimeter. A central table serves as a hub for socializing and lab meetings. A calendar hangs on the wall to let everyone know when people are out of town and when we'll be scanning people at the hospital. This gives a snapshot of how busy we are. No data, no science. I like to see a good flow of research subjects, with at least four a week. Other than that, the walls are covered floor to ceiling with whiteboards. We use the walls to graffiti ideas. Every inch is covered in diagrams, equations, or graphs. Visitors are mesmerized by the visual onslaught of the specialized code of science: Greek symbols, statistical arcana, flow charts. The lab people are literally surrounded by their ideas.

It takes about two years to go from initial brainstorming to published paper. The actual data collection — scanning subjects in the MRI machine — takes the smallest portion of that time. We might spend six months brainstorming and debugging an idea and only one month collecting the data. Sometimes the results turn out to be a lot more complicated than anticipated. Okay, *most of the time* they turn out to be more complicated than we had expected; sometimes we spend a year analyzing the data to make sense of the results. The process of writing up the findings and submitting to a journal to publish them can also take a year.

A few years before embarking on the Dog Project, my team began exploring different types of decision making. Having spent a decade studying the effects of rewards like money and food on the brain, we had recently branched out to study decisions based on sacred values. This was not planned. Instead, it came about when I met Scott Atran, an anthropologist who studies the roots of terrorism. We met at an academic conference and, over a bottle of wine, hatched the idea of using fMRI to try to understand how religion and other sacred beliefs guide decision making. It would be a fun collaboration, with the practical added benefit of being fundable by the Department of Defense. But in order to probe people's sacred values, we would have to push on hot-button issues. Race, religion, sex, guns, abortion, gay rights — all the stuff you don't talk about with the in-laws.

We spent a year brainstorming the sacred values experiment, at least half that time wasted because nobody in the lab truly felt comfortable talking about these issues. Scientists or not, if you really push on what is sacred to people, you can be sure they'll be offended.

At some point, I think the lab realized that we weren't going to make progress until we got better at suggesting ideas that might offend someone else. So it was with a determined effort that we became truly politically incorrect. That's also how we really got to know one another. The team includes people of different sexes, sexual orientations, religions, races, political affiliations, even diets. Drawing on our own sacred values, we each compiled a list of the most offensive statements we could imagine and whittled them down. When we examined the brain responses to these statements, we found that the brain processes sacred values as rules — like the Ten Commandments. This was important because it explained why sacred beliefs are so resistant to change. They cannot be argued with, and they cannot be traded for money or other material things.

Maybe it was some kind of cosmic premonition, but one of the issues we probed in the sacred values experiment was whether people identified themselves as a dog person or a cat person. I am not sure this is a good thing, but I have always categorized people this way. And if the answer was "neither," then that was the worst of all.

Against this backdrop of the sacred values experiment, the mission to kill Osama bin Laden was all over the news. As details trickled out, it was revealed that a dog had accompanied SEAL Team 6.

This shouldn't have been particularly surprising; dogs have been part of military units throughout the twentieth and twenty-first centuries. They are fixtures at border crossings and airports, and every urban police department has a K9 unit. But the fact that a dog had helped kill the most wanted man in the world was something special. It showed that dogs were not just companions. Even though it could have no understanding of democracy, a dog had helped defend a way of life.

Like the human members of SEAL Team 6, the identity of the dog on the mission wasn't revealed. But this anonymity just stoked the media firestorm. To satisfy the public's appetite for details, the public relations arm of the navy released stock photos of military working dogs: A German shepherd wearing a bulletproof vest bounding through a stream. A Belgian malinois, in tandem with its handler, leaping off the ramp of a helicopter.

The most touching photo was of a dog strapped to the chest of a soldier parachuting out of an airplane at thirty thousand feet, both wearing oxygen masks. The soldier cradled the dog with one arm while pulling the parachute release cord with the other. The closeness of the bond and the physical embrace really hit home for me: dogs and humans belong together. We couldn't exist without each other.

Prior to seeing those photos, I had been completely unaware that dogs had been trained to do such amazing feats. The noise from a helicopter is deafening. Most humans take some time to get used to it, and even then they wear heavy-duty ear protection. Obviously these dogs had been acclimated to some fairly hostile environments. Judging from the photos, they not only tolerated them, they enjoyed working in them with their humans.

"Did you guys hear there was a dog on the SEAL team?" I asked at our Wednesday lab meeting. The team came over to one of the computers to see the images of the military dogs that had gotten me so excited.

"That's badass!" Andrew Brooks, the sole graduate student in the lab, said. Andrew had been in the lab for two years and was working toward his PhD in neuroscience. I liked him a lot. His parents were missionaries then living in Japan. But their religious fervor didn't stick to Andrew. He swung the other way and found his calling in science. Even so, his route to Emory University was unusual.

Emory is considered a fairly prestigious institution, and most of the students who apply to graduate school come from a predictable group of universities. The Ivy Leaguers tend to stay in the Northeast, so Emory gets a steady feed of students from the "Southern Ivies" like Duke and Vanderbilt Universities. But Andrew had gone to a local community college and then transferred to a tiny liberal arts school in Macon, Georgia. After he graduated, he had applied to grad school at Emory. Macon is about as deep in the South as you can get. I knew Macon only as the home of the Allman Brothers Band and the place where Duane Allman was killed when his motorcycle collided with a flatbed truck in 1971, leading to the posthumous classic album *Eat a Peach*.

Early in my career, I would have turned up my nose at a student like Andrew. There was a time when I mistook pedigree, or even raw intellect, as the key determinant of success in science. But I had grown wary of the paper superstars. Too many incredibly smart students had come through the lab who didn't have the passion for research. Maybe they were accustomed to things being easy for them. Unfortunately, science never goes the way you expect. Many of them didn't deal with the unexpected very well.

Andrew didn't take anything for granted. He was smart, he worked hard, and he had a fire in the belly for doing experiments that might fail spectacularly. And Andrew was a dog person. He lived with a toy poodle named Daisy and an American Eskimo called Mochi.

The other big dog person in the lab was Lisa LaViers. Lisa had just joined the lab after graduating from Emory. She had done well in my neuroeconomics class the previous semester, and when a job opened up in the lab, I had encouraged her to apply.

Lisa was, in a word, perky. As one of the younger people in the lab, I loved her sense of adventure and the enthusiasm that she brought to the team. Although she had no previous experience with fMRI, I went on a gut instinct that she could quickly learn the skills to carry a project from the starting line all the way to the finish. She had majored in economics, so she had some math skills. Everything we did in the lab, from programming experiments to analyzing the fMRI data, involved a fair amount of mathematical sophistication. Even so, nothing an econ major couldn't handle. Although Lisa was initially apprehensive about taking a job for which she was a newbie, she quickly gained confidence as she took over the sacred values project.

Lisa's most endearing feature was what she referred to as her birth defect. It was more like a mannerism. Whenever Lisa listened intently to someone talking, she would wrinkle her eyebrows in a Spock-like expression. Most people interpreted this as a sign of confusion. Since Lisa was a social person who listened to a lot of people, some people concluded she was perpetually confused.

But Lisa was never confused when it came to dogs. She was head over heels in love with her two-year-old goldendoodle, Sheriff. Sheriff was a big, goofy dog. Larger than both a standard poodle and a golden retriever, he was imposing until he opened his mouth in a grin that broadcast, *I love you, whoever you are.*

After everyone had seen the pictures of the military dogs, the group settled in around the central table.

"If dogs can be trained to jump out of helicopters," I began, "then surely they can be trained to go into an MRI."

Andrew nodded. Lisa's eyebrows crinkled up.

Gavin Ekins was the first to ask the obvious question: "Why would you do that?"

Gavin had been in the lab for two years. After receiving his PhD in economics, he had joined the group to learn about the imaging side of neuroeconomics. I could always count on him to get right to the heart of the matter. He was dogless because of his living situation but had grown up with dogs. He was dating a girl whose role it was to assess the monkeys used in research at Emory for cage compatibility. A monkey matchmaker.

To Gavin's question, I replied, "To see what they're thinking."

"I don't think you need an MRI to do that," Gavin said. "It's 'Squirrel!'"

That got a good laugh—we were all fans of Pixar's *Up*—which of course triggered a round of other what-dogs-are-thinking jokes, centered around food and butt sniffing.

Monica Capra surprised me by being the first in the lab to say this was a good idea. Born and raised in Bolivia, a country ravaged by poor economic policies, Monica had obvious reasons for becoming a professor of economics herself. Unsatisfied with theory, she had gone on to specialize in experimental economics, doing actual tests to verify that people behaved the way other economists said they did. A colleague had introduced us eight years earlier, and because of our mutual interest in decision making, we had hit it off, designing fMRI experiments together ever since.

Monica was a tough cookie, always critical and not shy about poking holes in the ideas of others. Underneath her shell she was a warm person, but she was allergic to dogs. She was the last person in the lab I would have expected to support this.

"People spend an enormous amount of money on their dogs," she said. "They are important to many people. I think it's important to figure out why."

Kristina Blaine, who coordinated all the activities of the lab, voiced her support too, which was strange considering that she lived with four cats.

Sitting next to Monica was Jan Barton. Jan (pronounced *yahn*) is also from South America, in his case, Argentina. Jan is a professor of accounting. Monica had told him about the kind of research we were doing in the lab, and he had started hanging out with us to figure out how to use neuroimaging in accounting, which was a completely novel application of fMRI and something nobody had done before — always a risk to one's academic career. Jan had a dog that was on Prozac for anxiety — he just smiled at the idea of scanning dog brains.

Lisa had been deep in thought and said finally, "If we start scanning dogs, does that mean we'll have dogs in the lab?"

"I guess it does."

"Yaaayyy!"

I turned to Andrew. There was no way I was going to be able to do this by myself. I still had to teach and supervise the rest of the research projects in the lab. Andrew was the only grad student. This meant he had the most free time to spare. He was also the only person in the lab besides myself who had the necessary technical knowledge about MRI.

"Andrew, do you want to do this?"

"Hell, yeah!"

"Not to rain on the puppy parade," Lisa said, "but what is the scientific question?"

There are two types of experiments in science: fishing expeditions, where you start collecting data without a clear idea of what the right questions are, and hypothesis-driven experiments, where you start with a specific question to answer. Every middle school student would recognize the latter type as the foundation of the *scientific method*. Most people think that hypothesis-driven experiments are the only way scientific progress occurs. And science journals strongly prefer hypothesis-driven experiments.

The recipe for the typical hypothesis-driven experiment is simple: Take a well-accepted scientific theory. Find some minuscule aspect of that theory that nobody has ever verified before. Do an experiment that proves that aspect and supports the theory as a whole. Publish.

These experiments make for easy reading and are a surefire way to get results published, building up a résumé that will ensure promotion and tenure at a university. These types of experiments are also popular with funding agencies because the risk of failure is minimal. By my estimate, nearly all published research falls into this arena.

The thing is, hypothesis-driven experiments are incredibly dull. Most of the time you don't even need to read the experiment to know that the scientists have proven what they basically knew in the first place. If you already have a well-accepted hypothesis, then you already know the most interesting aspects of the scientific question, and the experimental results will, at best, advance knowledge incrementally. Of course, if the hypothesis turns out to be wrong, that would be really interesting. But those results are almost impossible to publish because nobody believes them.

In answer to Lisa's question, I said, "This is a fishing expedition. It is an idea in search of a question."

Andrew frowned, clearly troubled by the conflict this would cause with his dissertation research. The standard curriculum of any graduate program in science drills into students the importance of having a clear hypothesis for their research. But I had no hypothesis for the Dog Project. I had no idea how we were going to do this or how long it would take. Frankly, it probably wouldn't even work.

"Andrew," I said, "the Dog Project will be high risk. But it's going to be a blast, and I guarantee you that if it works, we'll be the first to have pulled it off."

"I'm in," he said. "But are we going to have to sedate the dogs?"

"Why would we do that? If they're sedated, then we won't know what they're thinking."

"So they'll be completely awake?" Lisa asked.

"They'll have to be," I replied. "Just like humans."

At the time, none of us realized just how much work lay ahead. We didn't know what the technical difficulties might be, considering dog brains are much smaller than human ones. We hadn't even begun to think about the actual experiments we might attempt.

At that point, it was all academic. Before we could go any further, we would have to figure out how to train a dog to go inside an MRI.

4

Puppy Steps

EVEN THOUGH CALLIE HAD BEEN in the house for a year, I had not completely warmed up to her.

I wasn't even sure that I liked her.

Kat knew how much I had loved Newton. When she and the girls went to the animal shelter, they had deliberately picked a dog that was about as different from a pug as you could get. Callie was the anti-pug. Pugs are short, stocky, and slow. Callie was a lean, mean fighting machine. Her muscles rippled beneath her thin coat.

Where Newton's face had been fixed in a permanent clownlike expression, Callie's was always on high alert. Her head was like a periscope, constantly swiveling back and forth in search of prey. Though she was quite friendly, her posture was off-putting to many of the dogs in the neighborhood.

Callie's strong drive caused endless distress to Helen and Maddy. Whenever Callie killed a chipmunk, the girls would berate her for her cruelty. To make matters worse, Callie wasn't cuddly. She didn't like to sit in laps. Sure, she would readily hop on the sofa, but then she would curl up like a cat at the other end — nearby, but not quite touching.

I missed my bedtime ritual with Newton. He would burrow under the covers, seeking refuge in my armpit, and I would pretend to protest. Although Callie wanted to sleep in the bed, her state of alertness never switched off. She would assume a position at the foot of the bed, facing the door, on watch for potential intruders or edible critters. Any attempt to move her unleashed a snarling, snapping bundle of fur. She wanted nothing to do with my armpit.

There was a dog-training facility in a strip mall within walking distance from our house. It was called Comprehensive Pet Therapy — CPT for short. Shortly after Kat adopted her, we signed Callie up for a basic obedience class.

CPT was the brainchild of Mark Spivak, who founded it in 1992. I first met Mark when we signed Lyra up for obedience training in 2005. Mark was not your typical dog trainer. He graduated from the University of Pennsylvania with a degree in economics and then received his MBA from the University of California, Berkeley. Mark bounced around the semiconductor industry in the Bay Area for a while but never meshed well with management. After he moved to Atlanta, he and his German shepherd, Topper, started competing in agility competitions to relieve some of his work stress. They did well, and Mark began helping friends with dog-training problems on the side. Within a few years, he decided to take the plunge and go into the dog-training business full-time.

Mark was a no-nonsense kind of guy. He employed several schools of thought about dog training, choosing the methods most appropriate for each dog and owner. And while he favored positive training methods, he acknowledged that punishment was also necessary from time to time.

Even though I hadn't yet bonded emotionally with Callie, I did enjoy working with her in Mark's obedience class. Lyra had taken this class too, but she had never had the level of intensity that Callie brought to the table. Callie wasn't warm and cuddly, but I had to respect her work ethic. She couldn't get enough training. She would do anything for a bit of hot dog. I was amazed that she learned basic commands like "sit," "stay," and "come" in just a few tries. The CPT teachers loved to use Callie as an example, because she watched them intently and worked tirelessly for a treat.

As Mark was the only dog trainer I knew, it made sense to approach him about the idea of training dogs to go into an MRI. He took an almost academic approach to dog training, so I hoped he would find the idea of scanning dogs' brains interesting enough to do for fun.

Much to my delight, Mark agreed to meet.

The modern study of dog behavior began with every biologist's hero, Charles Darwin. In *The Expression of the Emotions in Man and Animal*, Darwin devoted a great deal of attention to the dog—as an owner himself, his study of dog behavior didn't require a trip to the Galapagos Islands. What Darwin understood, and what every dog owner knows—but many research scientists seem to have forgotten—is that dogs have a rich set of expressions and body language. Darwin had no problem discerning joy, fear, and rage in dogs. He was primarily concerned with observing the expression of these emotions, not with the intent of training these intelligent animals, but rather to understand how human emotions evolved.

It was the famous Russian physiologist Ivan Pavlov who launched the modern era of dog training. Unlike Darwin, Pavlov had no love for dogs himself. He was just using them to study the digestive system. The problem was that his dogs started salivating before he fed them, and this messed up his data. Regardless of what you think about Pavlov, his "failed" experiment led to the most important discovery in psychology of the twentieth century, for which he was awarded the Nobel Prize in 1904. His discovery has completely dominated theories of dog training ever since.

Pavlov's discovery is called *classical conditioning* (although some people honor him by calling it *Pavlovian conditioning*). During the period in which Pavlov was doing his experiments, physiologists thought of the entire nervous system as a collection of reflexes, like the involuntary leg jerk when a doctor raps on your knee. They believed that all behaviors, even complex ones, were basically a series of reflexive actions. A reflex could be broken down into two parts: the unconditioned stimulus (US) and the unconditioned response (UR). For the knee reflex, the US is the hammer hitting the patellar tendon and the UR is the quadriceps contraction that results in the leg jerking upward. Pretty simple.

Pavlov realized that his dogs were having reflexive responses, but they weren't natural. Hungry dogs will always salivate when presented with food. This is a natural, and thus unconditioned, response. But, as Pavlov discovered, if something neutral, like the ringing of a bell, regularly precedes the presentation of the food, the dog will start salivating at the sound of the bell. The bell, a neutral stimulus, becomes a conditioned stimulus (CS), and the salivation it evokes is now a conditioned response (CR). The terminology of *unconditioned* and *conditioned* refers to stimuli and responses that are either natural or created by the experimenter.

By itself, classical conditioning doesn't say much about dog training. The responses are so simple that they don't constitute anything remotely resembling a behavior, and it is hard to imagine cobbling together a string of these conditioned responses into something as simple as "sit." This is where *instrumental learning* comes in.

In instrumental learning the animal must do a purposeful behavior. While classical conditioning trains an involuntary response like salivation, instrumental learning aims to train a voluntary action.

Instrumental learning forms the basis of every dog-training method ever published. Teaching the "sit" command is based on instrumental learning. Here, the stimulus is either a hand signal or a spoken word, and the desired behavior is the act of sitting. When the dog sits and he is immediately rewarded, he makes an association between the act and the reward. In instrumental learning, the link between stimulus ("sit") and act (sitting) is called the stimulus-response (S-R) relationship. Instrumental learning is also called operant conditioning because the animal learns to *operate* on, or affect, the environment.

Psychologists have classified four different types of instrumental learning based on whether a behavior is rewarded or punished. A reward is something that the animal likes, such as food or praise. Punishment is something he doesn't like, such as a loud noise. Rewards and punishments can be either given or withheld, which leads to the four types of learning. For example, the removal of something unpleasant reinforces behavior, so we call it negative reinforcement, *negative* meaning "removal." Positive reinforcement comes from the delivery of a reward, while positive punishment comes from the delivery of something unpleasant. The final combination, negative punishment, occurs when you take something desirable away from the animal. Negative punishment is a popular tactic among parents trying to curb undesirable behavior in their children. The suspension of computer privileges is a classic negative punishment and should, according to theory, decrease the frequency of the offending behavior.

The use of instrumental learning to change behavior is broadly referred to as *behaviorism*. Psychologist Edward Thorndike described many of its basic laws. The *law of effect* states that S-R relationships are determined by how much the animal likes the reward. The more he likes it, the stronger the S-R link. Thorndike's *law of exercise* states that an S-R relationship is strengthened through use and weakened through disuse. Thorndike's laws were further elaborated by the legendary psychologist B. F. Skinner, who thought that all behavior could be reduced to a set of S-R relationships. He is most famously associated with the Skinner box, a device that automatically trains rats or pigeons to learn behaviors.

After Pavlov's basic discovery and Thorndike's and Skinner's elaborations on it, behaviorism flourished. It reached its peak in popularity in the 1960s, when psychologists and psychiatrists began applying these theories of animal learning to human behavior. Techniques that targeted everything from smoking cessation to learning to make friends were all rooted in the behaviorist tradition. While some of its prominence has waned in recent years, behaviorist techniques remain the most commonly used "talk therapies" for depression and anxiety in humans, which is called *cognitive-behavioral therapy* (CBT).

When it comes to dogs, much has been said and written about positive and negative training methods. While they are all based in the behaviorist tradition, different schools of thought place different emphases on rewards like food and praise and punishments like noises, scolding, or pain. There is no doubt that the administration of a punishment can cause an immediate effect on a dog's behavior. What is unclear is whether the dog actually learns anything from it. The child who has lost her TV privileges may have learned not to repeat her offense, or she may simply have learned not to get caught.

This is the limitation of behaviorism: one can never truly know why a person or animal does something. You can only observe the effect of a reward or punishment and whether it increases or decreases a particular behavior. In fact, hard-core behaviorists completely dismiss what goes on in an animal's head. Since behavior is the only thing that matters to a behaviorist, subjective thoughts and emotions become irrelevant. But if you have tried to curb a dog from a particular bad behavior—chewing furniture or shoes, for example—you know the frustration of trying to understand why none of the punishments are working. How many dog owners have cried out in vain, "Why are you doing that?"

I hoped the Dog Project would someday be able to answer that question.

Until that day, Mark and I would need to figure out a training protocol based on conventional behaviorist methods that would get a dog to willingly climb in an MRI machine.

I met Mark at CPT. The training facilities are basically a large room. The linoleum flooring makes for easy cleanup of the inevitable "accidents." Apart from a teeter-totter and some ramps and hoops for agility training, the room is devoid of furniture. The spartan decor minimizes dog-induced damage expenses.

Mark was wearing his standard attire for dog training: a polo emblazoned with the CPT logo, athletic shorts, and running shoes. I had seen him only in dog-training mode, so I was surprised when he greeted me with such enthusiasm for the Dog Project.

From the beginning, we agreed training should be done strictly with positive reinforcement. It wouldn't be right to use punishment to teach a behavior this strange that would not directly benefit either the dogs or their owners. Everything in the Dog Project should be fun. Fun for the dogs, and fun for the owners. Mark suggested that this would be much easier if we could utilize the dogs' natural behaviors.

Natural behaviors are ones that dogs do on their own. Walking, sitting, and lying down are natural behaviors. If the dog has a drive to hunt small animals, then tracking might be considered a natural behavior too. Retrievers were originally bred to retrieve ducks, so they have a natural drive to carry objects in their mouths and, at least in theory, return them to their handlers. For some dogs, swimming is a natural behavior. For others, water is to be avoided at all costs.

It is safe to say that going into an MRI is not a natural dog behavior. Most humans don't like it either. But Mark explained how we could train a sequence of behaviors that were mostly natural for the dog.

"Most of what the dog has to do is a 'down-stay' position, correct?"

In a "down-stay," the dog lies down and stays in that position while the handler remains some distance away.

"Yes," I replied.

"Lying down is a natural behavior, so that is easy to teach with positive reinforcement. What else does the dog need to do?"

"He needs to hold his head perfectly still," I said.

"How still?"

"Less than two millimeters of movement for periods up to twenty seconds."

Everything depended on the head being still. Any movement would render the MRI data useless. When we perform scans on humans, the subject lies on her back with her head surrounded by foam pads. Most people are able to remain still, and the foam makes it easier. But a dog might not like his head being encased in foam. Maybe something less intrusive would suffice.

"We could make a chin rest for the dog," I suggested.

Mark liked this idea. "When we train dogs for tracking, we will often teach them a 'touch' command where they touch their nose to a target. We could do the same thing to teach a dog to 'touch' a chin rest."

Dogs use their nose to touch and sniff everything. This was a brilliant example of taking a natural behavior and turning it into a trained one. That left only the noise. MRIs are as loud as a jackhammer.

Mark stressed the importance of subject selection. He said, "We will need to carefully select the first subjects for the right tempera-ment characteristics." With the right subjects, the training would be easy. We certainly didn't want a situation in which the dog didn't want to be there. Even if we could train the dog to stay in the MRI, if he didn't want to be there all we would capture would be an anxious dog brain.

Because the patient table of the MRI is elevated, the ideal dog would be unafraid of heights, let alone enclosed spaces. Because we would most likely be studying several dogs, the ideal subjects would need to be social. And because there would be different people at the scanner — including MR techs, vet techs, and people from the lab — the dogs would also have to be unafraid of strangers.

In Georgia, thunderstorms occur with regularity during the spring and summer. I don't know if there is a higher proportion of dogs with thunder-phobias in the Southeast, but it is very common in Atlanta. Even though the MRI doesn't sound like thunder, an existing nega-tive association to loud noises might make training difficult. As long as the dog didn't have a noise phobia, we could gradually acclimate him to the specific type and volume of noise the MRI makes.

"The dog should be calm," Mark said. "And he should be unafraid of novel environments."

I had no grant funding to do this. Everyone was volunteering, but it still cost $500 an hour to rent time on the MRI. I had a small amount of discretionary research funds, but to keep costs down, we couldn't burn up scanner time just to let the dogs get used to the room. If we could find dogs that naturally remained calm in new situations, it would significantly boost our chances of success when it came time to actually do the scans.

"The most important trait," Mark said, "is motivational drive."

"What do you mean?" I asked.

"The dog has to enjoy the training. If he isn't having fun, it is much harder to shape behaviors."

Thorndike's first law. The more the dog likes something, the stronger the S-R relationship.

"Do you know any dogs that meet all these criteria?" I asked.

"I know some that compete in agility trials," he said. "But the owners can be a problem. If the owner isn't motivated to do the training, then the dog won't be either. A lot of people in the dog world have their own ideas about training. For this to work, we will want the training protocol to be consistent between dogs and owners."

I hadn't thought about the human side of the equation. Getting people to do what you want is a lot more difficult than getting dogs to. If Mark could do all the training, that would solve the problem, but he still had a business to run. What if I, or Andrew, learned how to train dogs? I wondered whether Callie could do this. She certainly wasn't calm. But she was highly motivated by the prospect of hot dogs. The idea of training Callie to go into the MRI seemed unlikely, so I kept that thought to myself.

Mark had been in dog training for a long time, and he knew a lot of dogs and owners in Atlanta.

"I have a few people in mind," he said. "Let me talk to them and get back to you."

I was excited. I didn't think anyone in the dog world would take the idea of scanning dogs' brains seriously. But Mark was not your typical trainer. Much to my surprise, he was just as excited about the Dog Project as I was. After twenty years of dog training, he was feeling a bit burned out. The Dog Project, he later told me, renewed his enthusiasm for his work, opening up a whole new dimension in improving dog-human communication.

5

The Scanner Dilemma

WHILE ANDREW AND I WERE pretty sure we could figure out how to scan a dog's brain, we had neglected to consider a minor, though important, detail: Where? The Dog Project needed a home.

The lab had been captivated with the "big question" — figuring out what goes on in a dog's brain. Details like the type of brain scanner, or where to find it, were just that: details. Up until this point, I hadn't been concerned. The best part of being a scientist is when the ideas are coming so fast and furious that you can't even write them down. You don't have time to worry about details. They just get in the way.

But eventually we had to confront the practical aspects of pulling this off. And the first detail was finding an MRI facility that would let us bring dogs into its scanner.

Yerkes National Primate Research Center, located about a mile from the main Emory campus, was our first choice for the MRI scanning. Nestled in a valley lined with southern pines, Yerkes seemed ideal. It was a short drive from the lab, so we could easily move our equipment there. And because it was off the main street, it was also quiet

and peaceful. The last thing we wanted was to scare a potential canine subject with a trip through a busy intersection. From a dog's
perspective, I imagined Yerkes would look like a walk in the woods.

Yerkes also specialized in the study of animals — primarily monkeys. Andrew and I congratulated each other on our good fortune.
We had come up with the idea of scanning the brain of a fully awake
dog, and one of the premier facilities for the study of animals turned
out to be right in our backyard. In fact, there are only eight such
facilities in the United States. Yerkes even had an MRI scanner dedi-
cated specifically to the study of animals. A friend and colleague of
mine, Leonard Howell, was director of the Yerkes Imaging Center
and invited us to take a look at how they scan monkeys' brains.

Although the Yerkes MRI center is unusual in the sense that it was
purposely built for the study of how primate brains function, it is
actually not that unusual to have such a facility at a veterinary school
or even at a high-tech veterinarian hospital. Any and all medical di-
agnostic tests performed on humans are now also done on animals.
The challenge with obtaining an MRI of an animal, however, is that
the subject must remain absolutely still. In a veterinary setting, this
means sedating the animal with medication. But sedating an animal
means that you can no longer study how the brain functions.

Leonard had pioneered a new approach to studying monkeys'
brains. Instead of sedating the monkeys, he had figured out how to
scan their brains while fully awake. This was a big deal to neuroscientists. When you administer drugs that render the subject unconscious, you change brain function in a major way. How this happens
is not really understood. While the unconscious state is interesting
for its own sake, most neuroscientists spend their time trying to figure out how the conscious brain works. Having conscious subjects,
animal or human, is critical.

Working with monkeys is a dangerous business. Monkeys are mean.

Not if-you-don't-give-me-food-I'll-ignore-you mean. More like if-you-don't-give-me-food-I-will-rip-it-from-your-hand-and-eat-your-finger-and-chew-off-your-face-for-dessert mean. This presents certain logistical problems for scanning their brains, especially if they are to remain fully awake.

What's more, because they are closely related to humans, diseases can pass between the species with ease. For instance, HIV, the virus that causes AIDS, is believed to have originated in African chimpanzees. Monkeys harbor a strain of the herpes virus that is fatal to humans, which can be passed along if, for example, one spits on you, which monkeys often do. The monkeys also have to be protected from us. If humans can catch diseases from monkeys, the opposite is also true. Monkeys are particularly susceptible to tuberculosis. For all of these reasons, scientists must take extraordinary safety precautions to work around monkeys.

Andrew and I made special arrangements to see how Leonard and his team scanned the brains of fully awake monkeys. After registering at the security desk, we were escorted through a series of keyed doors and deposited in a changing room.

"You need to gown up," Leonard's assistant instructed. "From this point forward, everyone must be fully protected. This means gown, face mask, and eye shield."

The so-called eye shields covered our faces entirely and were claustrophobic. They also had a tendency to fog up. The face masks were the surgical type. The combination of shield and mask made speech about as effective as talking into a pillow.

Our first stop was the training lab. Three oven-sized stainless steel boxes lined one wall. They resembled small refrigerators, but the hasp-type handle suggested something akin to a pottery kiln.

"These are the training boxes," the assistant said. Opening one revealed a sterile interior with white enameled walls and a cubby for devices allowing tubes and wires to snake out to various pieces of monitoring equipment.

On the other side of the room sat an upright tube constructed from PVC plumbing material. A foot in diameter and three feet tall, the top end was capped with clear Plexiglas. A four-inch slot was cut in the center of the cap, and a plastic shelf sat below the slot.

The assistant explained, "This is the restraint device. The monkey has a collar around its neck that fits into the slot. With its head poking through, it rests its chin on the shelf."

Andrew pointed to a pair of hoses that were attached to the bottom of the device. "What are these for?"

"Waste drainage."

Pushing the resulting image out of my mind, I asked, "How do you get the monkeys to go in there?"

The assistant pointed to a metal rod on the wall. "That affixes to their collar, and then we can steer them into the device from a safe distance."

So far, none of this was looking appropriate for the Dog Project. I kept silent, though, still eager to learn anything that might be useful for us. The device kept the monkey from escaping, but it wasn't clear what would keep its head still.

The assistant pulled a pink block of foam from a shelf.

"This is how we immobilize the head," he explained. "First, we make a mold of the monkey's head, which is then used to make a positive cast with plaster. From that, we use a gel-type material to make a soft cast, which fits snugly around its head. We cut holes for the eyes, nose, and mouth. This gets clamped to the restraint device."

"And the monkeys cooperate with this?" I asked.

"They learn," he replied. "We shape their behavior through rewards. It takes about six months to train a monkey to go into the restraint device."

"What are the boxes for?" Andrew asked.

"Those are conditioning boxes. Once the monkeys are trained to go into the restraint device, the whole rig is placed in the box. We then train them with lights and sounds."

"Trained for what?" I asked.

"To get addicted to drugs."

Right. Leonard's research group was studying the biology of drug addiction. To understand addiction, you need to look at the whole process, from the first time somebody uses a drug to the point he becomes addicted. Because it is unethical, obviously, to get people addicted to drugs, Leonard uses monkeys as a stand-in.

The assistant continued. "Once they are trained to associate cues with drugs, we take the whole rig to the MRI scanner so we can see what is going on in their brains while they are craving drugs. Are you ready to go down to the scanner?"

I couldn't wait to get out of there.

Because the MRI's strong magnetic field affects computer equipment, the control room is partitioned from the main scanner room. When we entered, a young woman draped in a surgical gown was staring intently at a computer screen with several brain images.

She was not pleased to have visitors.

"Who are you?" she snapped at me. "Have you had a TB test?"

I honestly couldn't remember when I had last been tested for tuberculosis. Fortunately, Andrew distracted her.

"I have!" he announced cheerfully.

Leonard's assistant explained that we were there to observe MRI scans of monkeys. The monkeys being scanned that particular day were from a different research lab. Because they had not gone through Leonard's behavioral training, these monkeys had received a heavy dose of sedation. One monkey, surrounded by three veterinary technicians, was in the scanner when we entered, attached to monitors that reported vital signs like heart rate, breathing, and body temperature. Another monkey was on a cart, recovering from anesthesia. I almost walked right by it, until it started twitching with muscle spasms as the sedation wore off.

We took the opportunity to explain what we were trying to do with the Dog Project. The vet techs were not enthusiastic.

"You're going to have to monitor them," one said. "Vital signs and core body temperature."

"How do you do that?" Andrew asked.

"Rectal probe."

"Why would we do that to a dog that isn't even sedated?" I asked.

"It's standard operating policy to fully monitor all animals undergoing a procedure," she replied.

"But we're not doing a procedure," I protested. "The dogs will be trained to go into the scanner willingly."

She wasn't buying it. "Who is going to be with the dogs?"

"Us, the dog trainer, and the owner."

She shook her head. "I suppose you two are okay because you're university employees, but no outside visitors."

Although it was clear there was no convincing this woman, I pressed on. "Look, would you volunteer your dog to be in an experiment without being present?"

"I suppose not. Even so, you'll have to convince the review committees."

Andrew and I had seen enough. It surprised me that one of the nation's premier animal research facilities wasn't more encouraging about the Dog Project. But we were more determined than ever to find the right home for it.

When I got home that night, Callie and Lyra greeted me with unusual attention. Instead of jumping up and down as they usually did, they sniffed my feet intently. As I walked through the house they trailed me from a respectable distance, focused on my feet.

They knew. I had tracked monkey stink home with me.

Logistical problems aside, I realized there was no way we could do the scanning at Yerkes with all those monkeys.

6

Resonant Dogs

WHEN HELEN AND MADDY started kindergarten, I began a tradition of visiting their classes every year to teach the kids about the brain and perhaps convey some of the excitement in figuring out how it works. The first time I did "Brain Day" at the school, the principal and I had a frank discussion of what I planned to discuss.

"Will you emphasize the importance of brain health?" she asked. "Tell the kids about wearing bike helmets and how drugs damage the brain?"

"Um, sure," I said. "How do you feel about me bringing a brain to school?"

"You mean a plastic model?"

"No. A preserved human brain."

"In a jar?" she asked.

"A bucket," I explained. "We have a set of teaching brains at the university that I can check out. The kids can touch it."

A look of fascination flashed across the principal's face, immediately replaced with one of consternation.

"We'll need to send home a permission slip."

She needn't have worried. Not a single parent objected.

The kids loved Brain Day. Even a few teachers snuck into the classroom to touch the brain. I'm not sure the students ended up remembering much of what I said that first time, but it certainly made an impression when I reached into the bucket and brought out a full-sized, dripping wet human brain. Half the class said, "Cool!" while the other half simultaneously said, "Gross!"

By the time of the Dog Project, I had done Brain Day seven years in a row. Maddy was in fifth grade, her final year in elementary school, and Helen had begun middle school. The questions the students asked always fell into a predictable pattern. The bright ones asked questions like "Where do dreams and emotions come from?" Others just wanted to jam their fingers as far into the brain as they could. The last year I did Brain Day at the elementary school, a small boy raised his hand and asked a question I had never heard before.

"Have you ever studied a dog's brain?" he asked.

The teacher chided the boy for asking silly questions.

"As a matter of fact," I interrupted, startled by the coincidence. "We are about to do just that."

With Helen's transition to middle school, there wouldn't be an opportunity to bring the brains to her science class. Sixth-grade science was devoted to geology, meteorology, and astronomy, and biology wouldn't return until the seventh grade.

Growing up, Kat and I had gone to public schools, and we believed strongly in public education. As is true in many cities, however, the quality of the public schools in Atlanta varies widely. The schools that Helen and Maddy attended were solid but had the difficult mission of fulfilling the needs of all the kids in a very diverse district. A large number of children couldn't afford to buy lunch and many had special needs.

At the end of her first week of classes in middle school, Helen brought home her science textbook, one apparently compiled by a team of bureaucrats who had overdosed on their daily Ritalin. Every page was crammed with full-color pictures guaranteed to distract even the most focused student from the text. The text itself was nothing more than a litany of facts to be memorized. Although it was the neighboring school district that had made national headlines for banning the word *evolution* from its textbooks, you could still detect a patronizing tone throughout. More than anything, it smacked of scientists-say-it-is-so (wink-wink).

Helen struggled. Although she was diligent with her homework, her test and quiz scores hovered in the mid-70s. Kat and I didn't want to be helicopter parents, but we couldn't let Helen flounder. It was time for a parent-teacher conference.

Helen's science teacher was a pleasant man who bore a striking resemblance to Ed Helms. The classroom looked much like I'd expected it to: slate laboratory tables arranged in neat rows, a chemical sink with an eyewash station should any mishap occur, wall cabinets full of rock specimens, a large periodic table of the elements on the wall.

After an exchange of pleasantries, I moved on to the reason for our meeting. "We're concerned about how Helen is doing in science."

He pulled up a grade spreadsheet to show us.

"Helen's a good student," he said. "She turns in all of her homework."

"Yes," I said, "but she seems unclear on what material she will be expected to know."

"The students get exposed to the material multiple times," he explained. "They hear about it in class. They read it in the textbook. And then we review it."

This may have been partly true, but having helped Helen with her homework and then heard what was on each test, I was skeptical. Helen was in fifth-period science, and I began to suspect that the teacher might have been confusing what he had gone over with the classes at the beginning of the day with those at the end.

"Helen said her class is noisy and that she has a hard time hearing what you're saying."

"By fifth period," he replied, "the kids have a hard time sitting still."

Kat and I had already heard about his method of making the kids walk laps around the hallway to burn off energy. Maybe this helped some students concentrate, but it took valuable time away from Helen actually learning science.

"Can we move her to a different period?" I asked.

"We can check, but that would require changing her whole schedule."

"Can you at least move her to the front of the class so she can hear better?"

I think he realized that this was the least painful way to get rid of us.

"Sure, I can do that."

It was evident that he had been through this type of conference countless times before and that he had heard it all. I felt some small victory in serving notice that we cared about our daughter and that we would not sit idly while she slipped through the cracks of the public school system.

When we got home, Helen was in her room doing homework. I sat with her on her bed. Lyra jumped up to join us.

"How did it go?" she asked.

"Not so good," I said.

A look of embarrassment flashed across Helen's face. "What did you do?"

"We tried to get you switched to a different period, but that wasn't going to happen. The best we could do was getting you moved closer to the front of the room."

Helen nodded and stroked Lyra's head. Lyra grinned in delight.

"I think he forgets to teach your period some of the material," I explained. "You're just going to have to make a lot of flash cards."

Science is about questioning how the universe works and discovering new things, not memorizing a series of facts out of a textbook. Science constantly changes as we learn more about the world we live in. What could be more exciting than that? It saddened me that Helen had to learn science with all the life sucked out of it.

Helen continued to smooth out Lyra's fur.

"Do you think Lyra knows how I feel?" she asked.

"I think she does," I said. "But hopefully we can prove that through the Dog Project."

Lyra provided a great deal of comfort to Helen. As the two of them cuddled together, I was struck by their perfect symbiosis. As a golden retriever, Lyra had been honed through generations of selective breeding to get along with humans, especially children. Although the Dog Project had been conceived as an effort to discover what dogs like Lyra and Callie were thinking, Helen's reaction reminded me that the dog-human relationship is a two-way street. We couldn't consider the dog brain without taking into account dogs' effect on humans.

At a superficial level, you can state the obvious: humans like dogs. They provide companionship. They serve as working and utility animals. They hunt. They guard. They are soft and warm and feel good against the skin. But, as I was trying to convey to Helen, science is about asking why things are the way they are.

The scientific study of dogs' effect on humans has been, until recently, almost nonexistent. Florence Nightingale, the matriarch of nursing, was one of the first to argue for the role of animals in improving human health, writing, "A small pet animal is often an excellent companion for the sick, for long chronic cases especially." But it wasn't until the last decade, when animal-assisted therapy became more accepted as a treatment for human illness, that researchers began measuring the effect of dogs on humans. Even so, the results have been mixed. For one thing, how can you conduct a double-blind study where neither the researcher nor the patient knows what treatment is being given, if one set of patients gets to play with dogs while the other doesn't? Double-blind studies are the gold standard in medicine because of the well-known placebo effect. Across the board, for physical and mental illnesses, up to one-third of patients will get better if they believe the treatment they are receiving is effective, even if it is nothing more than a sugar pill.

Demonstrating that dogs and animals in general can improve human health probably won't meet most medical standards of evidence. But that doesn't mean animals don't help people. One study found that animal therapy helped hospitalized heart failure patients by decreasing blood pressure in the lungs, a measure of how much fluid is backing up. Another study suggested that animal therapy reduced the need for pain medications. Hospitalized children in particular seem to benefit from pet therapy, with marked decreases in pain experienced. Many of these studies, however, have used subjective measures like pain as their endpoints. The few studies that have attempted to measure the effects of animals on human biologic measures, like blood pressure or stress hormone levels, have come up with contradictory results.

Interestingly, when you look at the entire literature on animal-assisted therapy, patterns begin to emerge. Of the different animals used in therapy, dogs are the ones associated with the largest beneficial effects on health. And although positive effects were observed in most age groups, children seem to derive the most benefit.

Up until that point, I hadn't given much thought to how dogs and humans were matched to each other. But watching Helen and Lyra together, it became obvious that Lyra helped soothe Helen's frustration and that Lyra enjoyed doing so, curling up next to Helen when she was needed most. Callie was a different story. She wasn't nearly as demonstrative. Even her body language was different. While Lyra was content to put her head in Helen's lap, Callie preferred to curl up nearby, just out of physical contact. Lyra appeared to be well matched to Helen's personality, but it surprised me that Callie was better suited to mine. I didn't care for dogs that fawned over you like slobbering sycophants. I liked dogs that saw themselves as your partner.

In his book *Man Meets Dog*, the great Austrian ethologist Konrad Lorenz wrote about the different types of dog-human relationships. Lorenz realized that the loyalty of dogs had no counterpart in human relationships, but that alone did not make them better than people. He believed dogs are "amoral," without any instinctive sense of right and wrong. Modern research has disputed that statement. For instance, research by primatologist Frans de Waal shows that many animals demonstrate an understanding of fairness.

Lorenz, however, believed that the ideal canine companion was a "resonance dog." He noted the extraordinary parallelism in personality between many dogs and their owners, sometimes to the point that they even looked alike. According to Lorenz, strong dog-human bonds were created when both human and dog resonated with each other.

Certainly Helen and Lyra resonated. And even though Callie was the relatively new, and somewhat standoffish, dog in the house, I had to admit that she was beginning to resonate with me.

Leaving Helen and Lyra alone after our discussion about science class, I padded downstairs to find my resonant dog. As usual, she was in the backyard.

"Callie, here girl!"

She came bounding into the kitchen smelling like dirt and dog sweat. Wagging her stiff tail very quickly, she looked at me and ran out the door again. Clearly, she wanted me to follow her.

Callie had her nose buried in the ivy with her butt in the air. As I approached, she looked up and started shaking her rear end back and forth. Callie took something in her mouth and flipped it in the air. Whatever it was (probably a mole) emitted a high-pitched squeal, which was soon cut short.

I was impressed with Callie's hunting skills. Since she had no interest in eating her prey, she hunted either for her own enjoyment or for mine. There was no need to tell Helen about Callie's predatory activity. That would remain a secret between us.

"Good girl," I said. "You're a SuperFeist."

7

Lawyers Get Involved

DECADES AGO, WHEN COLLEGES had fewer rules and regulations, dogs were a fixture on campuses. Most fraternities had a dog, and professors often brought their dogs to class. What would college be without a dog chasing a Frisbee on the quad?

Sadly, those days are long gone. They were gone even before I went to college. But it's even worse now. Not only are dogs nowhere to be seen on a campus, but most universities explicitly forbid them. Only a handful of colleges, including Caltech and MIT, allow pets, and those are cats only. Lehigh University in Pennsylvania allows one cat or dog per fraternity or sorority, but it must remain in the house at all times. The only universities that allow dogs a relatively free run of campus are Stetson University and Eckerd College in Florida, the University of Illinois at Urbana-Champaign, and Washington and Jefferson College in Pennsylvania (although in the latter case you need to prove that it is a family dog you have owned for at least a year).

Emory University is not pet-friendly. Its policy manual reads, "Because of restrictions governing University insurance policies, concerns for the integrity of research projects, and interest in the welfare of faculty, students, staff, and visitors, it is the policy of Emory University that animals not be permitted in University buildings."

Insurance policies? Is the welfare of the university really threatened by dogs? It sounded like some lawyer had visions of a rabid dog running amok.

There was, however, a loophole in the prohibition against dogs: animals used for research were allowed. But this meant that several committees, buttressed with lawyers, would need to sign off on it. I mentally prepared myself to do battle against an army of "No."

When we do brain-imaging experiments on humans, all the procedures must be reviewed by a panel intended to protect volunteers from harm. Having performed close to a thousand MRIs on human subjects over the last decade, I have become accustomed to the approval process. But this time was different. We were going to scan dogs.

It is one of the sad facts of biomedical science that the road to scientific progress is littered with the bodies of both humans and animals. The modern era of human experimentation began with the Nazis. Doctors and scientists performed horrific experiments on people held in concentration camps, and all of this was justified in the name of scientific progress. During the Nuremberg Trials, these atrocities came to light. As a result, a code of conduct was established for how to do medical research without putting people at great risk. These rules have evolved over time, especially after some horrible lapses in judgment, like the Tuskegee syphilis study, which ran from 1932 all the way to 1972. In the Tuskegee study, researchers withheld treatment to disadvantaged African Americans without their knowledge so that the scientists could document the natural course of syphilis. In 1974, after the study was shut down, the National Research Act established a commission for the protection of research subjects. The commission produced a landmark document called *The Belmont Report*, which not only summarized the history of human biomedical experimentation since World War II, but also laid down guidelines for experiments involving humans.

The Belmont Report contains three basic principles for human research. First, there must be "respect for persons." This means that every person has the right to make his or her own decisions, including whether to volunteer for research. In other words, you cannot force or trick someone into participating in an experiment. Second is the principle of beneficence, which means that we should maximize the good resulting from research. This also means we should not harm people in the name of research. There is a bit of a gray zone here, however, in the recognition that all research carries some risk. As long as the potential benefits outweigh the risks, then the research is generally considered okay. For example, an experimental drug for cancer may have terrible side effects, but if it has the potential to save the subject's life, then the benefits might outweigh the risks. Finally, there is the principle of justice, or fairness, which means that scientists cannot use just poor people for research, because this would unfairly take advantage of their need to make money by renting their bodies for medical research.

While it has taken decades to work out how these principles are applied in practice for humans, the situation is entirely different with animals. The law does not recognize animals as having the same rights as humans. Legally, animals are considered property. This means that researchers can, within limits, do whatever they want with them. Usually, this means the death of the animal.

As bad as that sounds, the care of laboratory animals is highly regulated by the Department of Agriculture. The Animal Welfare Act, signed into law in 1966, specifies how animals used in research should be treated. Periodically updated, the text of the law is a mind-boggling list of rules that describes everything from cage requirements to veterinary care to methods of euthanasia.

The act requires any entity that performs research on animals, such as a university, to establish a committee to review and approve research protocols. This committee is called the Institutional Animal Care and Use Committee, or IACUC. The acronym is usually pronounced *eye-a-kuk*.

Let's say I wanted to conduct some behavioral research on Callie at home, like figuring out the best method to get her to come when called. As long as I didn't violate any animal cruelty laws, I could do whatever I wanted. Use a long leash? Fine. Try an ultrasonic whistle? Check. Use an electronic shock collar? Still okay. I wouldn't need anyone's permission to do any of this.

But if I asked the same question in an academic setting, like the university, it would fall under the legal jurisdiction of the Animal Welfare Act. If I wanted to write an academic paper on which dog biscuit was most effective for training, I would still need to get IACUC approval. The main difference between doing research at home and at the university is that the university is considered a "research facility" that receives money from the federal government. As part of the deal for receiving federal funds, the university must abide by all federal rules and regulations. One big part is compliance with the Animal Welfare Act. (The other is compliance with human research regulations like the ones established in *The Belmont Report*.)

Although I was accustomed to navigating the maze of human research rules, I had had no experience with the animal rules. Surprisingly, the rules of animal research were a lot more complicated. Unlike humans, animals have no choice in whether they want to participate in research. So while a human can theoretically judge the risks and benefits and make an informed decision, animals cannot. As a result, the rules around animal research acknowledge that their lives will be awful and limit as much as is practical the pain and suffering they must endure.

None of this seemed terribly relevant to the Dog Project. After all, the dogs were going to be people's pets. They weren't going to be housed at the university. The plan was for the owners to train their dogs at home and, when they were ready, bring them in for an MRI scan. Andrew and I figured this should be pretty simple. We wrote a document describing our plan for the experiment. This document spelled out the research protocol. It contained everything from how we would select the subjects, to how we would train them, to how we would protect their hearing during the scans. It even included a consent form (for the owner, not the dog).

We sent the protocol document to the IACUC and waited for a response.

Two weeks later, I received a phone call from a university lawyer.

"We have a jurisdiction problem with your protocol."

Trying not to get upset, I asked him to explain the problem.

"For starters," he continued, "you included a consent form."

"Yes," I replied. "We thought it was reasonable to get consent from the dog's owner."

"The IACUC doesn't do consent forms," he said. "This sounds like human research."

"But the humans aren't the subjects," I said. "The dogs are."

"Well, we don't know what to do with a consent form," he said. "You need to send it to the IRB." The Institutional Review Board, or IRB, was the committee that reviewed human research protocols.

"They're not going to want to review it because it's not human research."

"There are other problems," the lawyer continued, ignoring me. He then ticked off a laundry list of issues. Once on campus, how would we transport dogs to the MRI? How would we prevent the dogs from escaping? What would happen if they bit someone? The hospital risk management lawyers would have to sign off on this too. I would need to check with the Occupational Safety and Health Office to see if there were OSHA issues to resolve. I would also need to check with the biosafety officer to see if she had concerns about the spread of biological pathogens.

I couldn't believe what I was hearing. Suddenly, the little feist that slept in our bed and licked my face every morning represented a threat to the safety and welfare of the entire university.

"Have you considered purpose-bred dogs?" the lawyer asked, referring to dogs, usually beagles, bred and sold exclusively for research. There was no way I would support that ugly practice, and I said so.

"That would mitigate some of the liability concerns because Emory would own the dogs," the lawyer continued.

"We need to find a way to do this project with community-owned dogs," I said. "I'm confident that people will volunteer their dogs just to have a chance to participate in this research." Then I had an idea. "Do you have a dog?"

"Yes."

"Then surely you've wondered what your dog is thinking," I said. "Would you volunteer him?"

"Well, I don't think he would be a good subject," the lawyer replied. "But I see your point." He paused and then continued. "Maybe the IRB would act as a consultant to help us with your consent form."

A glimmer of hope.

"But because of the liabilities, you're still going to need all the offices to approve your protocol."

This was not going to be easy. I had interacted with some of these offices before, and I knew that nobody would want to be the guy who approved the crazy dog experiment. What if something went wrong? But there was no turning back. If I had to, I would do this off campus, on my own time. I would find some private MRI facility willing to take dogs.

One way or another, the Dog Project was going to happen, even if I had to fight every lawyer in Atlanta.

Many of the people who work in the divisions of the university concerned with regulatory compliance adopt a cover-your-ass attitude. This typically manifests as a preoccupation with the letter of the law. Unfortunately, there is an endless array of federal regulations, and they are not always consistent with one another, so knowing which rules take precedence in a given situation is a bit of an art. In my experience, many of the people in the compliance divisions were primarily concerned with minimizing the chance of any violation or anything that might bring negative publicity if something went wrong, without much regard to the potential benefits of taking that risk.

I called Sarah Putney, the director of the IRB. Sarah had always helped me work through ethical issues in our human work. She had an incredible knowledge of the rules, she loved research, and, perhaps most important, she was a dog person.

I explained what we wanted to do and Sarah immediately seemed to understand.

Getting right to the heart of the matter, she asked, "Who is the subject of the research?"

"The dog."

"Then this isn't something the IRB would review," she replied. "We only review human research."

"But we have a consent form," I explained.

"Why?"

I explained that since we were asking people to volunteer their pets for research, it seemed appropriate to explain what we were doing and what the risks were.

All research entails risk. In human research, the spectrum ranges from minimal risk to high. Minimal risk means that the probability and magnitude of harm in the research aren't greater than what is ordinarily encountered in daily life or during the performance of a routine physical or psychological exam. Anything more than that is considered moderate or high risk. But that is a judgment made by the IRB.

Our human fMRI work is considered minimal risk because we study normal, healthy people, and we don't give them any drugs. MRI doesn't use radiation, so it's considered very safe in and of itself. The main risks to humans are anxiety, because of the tightly enclosed space, and hearing loss from the noise. To limit the risk of a claustrophobic panic attack, subjects are given a button they can press if they want to be removed from the scanner. To protect their hearing, they wear earplugs and earmuffs.

The risks for dogs would be the same, in theory. Dogs have more sensitive hearing, so there might be a greater risk for hearing damage. To minimize that risk, the dogs would need to be trained to wear earmuffs, but I felt that their owners should be aware of all the things that might go wrong, however unlikely. In my opinion, the worst that could happen would be a dog escaping, getting lost or injured in the process.

Because what we were proposing did not meet the definition of human research, no federal law required us to have a consent form. But, as I explained to Sarah, it seemed like the right thing to do. We were about to make a decision that elevated the rights of dogs to the same level as those of our human subjects.

• • •

Ever since I started running a research laboratory, I have operated under a simple ethical principle: *Do not do any experiments that you wouldn't be willing to do on yourself or a loved one.* This is not a universally shared philosophy. Many scientists do experiments that they would never volunteer for themselves. There is no rule that says you have to. Everyone has their own opinion of the risks and benefits of volunteering for research. The principle of "respect for persons" allows everyone to make their own informed decision to participate in an experiment, including the person running the experiment. But what message would it send if I were unwilling to be a subject in my own experiments? I have had about fifty MRIs over the years. I have no qualms about crawling into one. I would put my kids in an MRI. And my dogs.

After explaining my logic to Sarah, we agreed that the rules governing research on children provided the best model for what we wanted to do. If adults want to participate in research, they simply need to understand the risks and benefits and make an informed decision. Children are different. Not only do they not have the legal standing to make their own decisions, but the rules also recognize that they don't have the necessary knowledge or experience to understand the risks and benefits.

Research on children is given special scrutiny. If the research is considered minimal risk, then the approval process is pretty much the same as for adults. The main difference is that the parent gives permission and signs the consent form. However, the child must still indicate a willingness to participate, which is called *assent*. If the research is more than minimal risk, several different factors are weighed, including the relative risk and likely benefit to the child.

Like our human studies, the Dog Project would qualify as minimal risk. So we simply copied a consent form that we had used in one of our previous fMRI studies of children. Wherever the words *your child* appeared, we replaced them with *your dog*.

That left the dog itself. Since a dog can't sign a form, how would we detect the canine equivalent of assent? With a child, assent is usually determined by asking the child. If he is old enough, he can sign a document of assent, which is a child-friendly version of the consent form. But if he is too young to understand or to express himself, the researcher must rely on his behavior. For example, if a mother signs up for a research project with her infant and the baby shows obvious signs of distress, like inconsolable crying, the researcher should interpret this as a sign that the baby doesn't want to participate, and the experiment should be stopped.

We could do the same thing with dogs and treat them like infant research subjects. If they showed any signs of not wanting to participate, we would stop the experiment. The simplest way to do this would be to dispense with restraints. If a dog didn't want to be in the MRI scanner anymore, he could simply get out. Same as a human.

Never mind that all previous animal research treated animals as property. Elevating the rights of a dog to that of a human child made both ethical and scientific sense. It was the right thing to do and it would result in better-quality data too. If the dog didn't want to be in the scanner, the data would be useless anyway. And if he were strapped down, we might not even know that he didn't want to be there.

With the ethics squared away, it was time to go back to the IACUC with our proposal. The committee was happy with the consent form and pleased that Sarah Putney had helped draft the language. That left just a lineup of university administrators ready to say no.

First up was risk management. The sole purpose of this department is to minimize the chance of anything bad happening. The risk management department looks at research starting with the worst-case scenario. What is the most catastrophic thing that could happen, how would that damage the university's reputation, and what would it cost the university to defend itself in court?

Although we said "dog," I think risk management heard "Cujo," after Stephen King's fictional rabid Saint Bernard. Worst-case scenario: dog escapes, runs amok on campus, bites student, student gets rabies and dies. Because we would not be scanning at Yerkes, the dogs would have to be under control at all times, particularly when they were being transported to the MRI scanner at the Emory University Hospital. Fortunately, the MRI room had a door leading directly to the outside of the building. We could walk the dogs right up to the door without having to go through crowded corridors. Risk management liked that. It minimized the chance of inadvertent human contact. On the inside, there were three doors between the MRI room and the rest of the hospital, so there was no chance a dog was going to escape once inside the scanner room.

Next up was employee health. Working with animals created an occupational hazard. What if somebody on the research team was allergic to dogs? Could the dogs leave dander in the MRI that would cause a reaction in someone who went in the scanner afterward? The solution to the first question was for all members of the lab to certify that they were not allergic to dogs. Most of the team had dogs, so this wasn't a problem. To deal with the dog dander, we would dispose of the linens and wipe down the scanner with disinfectant.

The only remaining hurdle was the biosafety office. The biosafety officer was also concerned about rabies. She suggested that the research team receive preventive rabies vaccinations. Never mind that there have been no cases of human rabies in the United States resulting from a domestic dog bite in the last decade. Because all the volunteer dogs in our study would have proof of vaccination, the chances of contracting rabies from them would be essentially zero. The risks from the vaccine were far greater. The preventive rabies vaccination requires three shots over a month, and 50 to 75 percent of people receiving it get mild to moderate side effects, including headaches, nausea, dizziness, abdominal pain, and fever. No thanks. After being confronted with this data from the CDC, the biosafety office backed down and agreed that the risks from vaccination were greater than the risks of contracting rabies from a domestic dog.

With the final administrative hurdle cleared, the army of lawyers signed off on the Dog Project. We received approval to initially study up to ten dogs. If things went well, we could seek approval for more dogs after that.

Now we just had to find some subjects.

8

The Simulator

ITH THE ISSUE OF where to scan the dogs out of the way, we could turn our attention to acclimating the dogs to the MRI. This meant that we would have to build an MRI simulator.

The entire project hinged on a dog's ability to hold its head still while in the scanner. But training a dog to hold its head still was going to be the easy part. Doing it in an MRI was a different story. The interior of the MRI scanner is six feet long and less than two feet in diameter. Many people don't like being stuck in a coffin-sized tube. Fortunately, dogs aren't like humans, and many breeds actually like to be in small spaces. Callie, being of the terrier family, had no such fears and loved to tunnel under ivy and into holes. Even so, all the dogs that were going to participate in the study would need to be trained to go into a tube of the exact dimensions of the MRI. Once they were acclimated to the tube, they would then have to be trained to put their head into the head coil. The coil, which is nicknamed the "birdcage" because of the resemblance, is even smaller than the MRI tube. The dog would need to shimmy its body and head into the center of the birdcage.

The second, more difficult aspect of the MRI was the noise. MRIs are loud. When performing a functional scan, the MRI sounds like a machine gun. At nearly 100 decibels, being in an MRI is like standing next to a leaf blower. Although not terribly painful for humans, dogs have more sensitive hearing, and we worried about damaging their ears. Also, many dogs are just plain afraid of loud noises. It would do no good to scan the brain of an anxious, frightened dog. Not only would we need to find dogs with a calm temperament, we would still need to get them used to the noise. The simulator had to mimic this key aspect of the MRI procedure.

It is commonly believed that dogs have more sensitive hearing than humans. But how much more? At the low end, humans can hear frequencies of about 8 Hz, which sounds like a very deep vibration. The high end is limited to about 20,000 Hz. This range tends to shrink with age, with the high frequencies being lost to ear damage from loud noises over the years. Most human speech is in the 300 to 3,000 Hz range. The first investigation of dogs' hearing was done in the 1940s. But because sound-generation technology was limited at the time, the scientists couldn't generate very high frequencies, and they couldn't determine the dogs' upper frequency range. It wasn't until the 1980s that technology could reliably generate high-frequency tones. In the 1990s, an even more sophisticated technique was developed. This technique measures electrical responses in the part of the dog's brain that responds to sound. These *brainstem auditory evoked responses*, or BAERs, are also used in humans. Today, most scientists agree that dogs can hear tones up to about 60,000 Hz, well beyond the human range.

The actual MRI scanner makes a wide variety of sounds. These noises originate from what are called the *gradient magnets*. There are two types of magnetic fields in an MRI. The main field is produced by the miles of superconducting wire that are wrapped around the bore. The main field never changes and is always on. The gradients are much smaller magnetic fields that are constantly changing during a scan. By switching the gradients on and off, we can select specific locations in the brain. A gradient can be switched on by running electrical current through it, which activates the magnetic field. The sudden inrush of electricity causes the magnet to expand slightly, and this rapid expansion causes a pressure wave inside the MRI, which we hear as a loud banging. The exact noise that it makes depends on the type of scan being performed and whether it is a structural or a functional MRI.

We still needed a physical mockup of the scanner to train the dogs in. Many universities that house MRIs for brain research have mock scanners. There are many situations where it makes sense to train people in the scanner before actually scanning their brains. Brain-imaging studies of children, for example, must first teach kids to lie still in the scanner by using a simulator. Because the MRI can be frightening, it is very helpful to allow children to get used to the environment before they go into the real scanner.

It isn't surprising that a few companies have cropped up to sell mock scanners. The price tag, however, is steep. When we embarked on the Dog Project, the going rate for a mock scanner was about $40,000. Since we had no funding, this wouldn't be practical. Besides, I couldn't see spending so much money for what amounted to an empty tube with a few speakers inside to simulate the noise.

But how much of the real MRI would we need to simulate? Did we need something that could fill an entire room, like the real scanner? Or could we get away with a simple tube? After all, the dog was going to be inside only the scanner.

Mark came to the hospital to check out the MRI facility and determine how much of the actual MRI we would need to simulate for dog training. He hopped up on the patient table and lay down, putting his head in the birdcage. With the press of a button, the table glided into the center of the magnet. A quick thumbs-up indicated he was good to go. We went through a few quick brain scans so that Mark could get an idea of the types of noises and how loud they were.

Mark came out of the scanner with a big grin on his face and proclaimed, "This is completely doable."

"Do you think we need a mockup of the entire scanner?" I asked.

"No, just the patient table and the tube will be fine."

This was good news, because we were going to have to build the simulator ourselves. The mock scanner would have three elements: a tube to simulate the inside of the scanner; an exact model of the birdcage, which the dog would have to shimmy into; and a sound system to play recordings of the scanner noise at the appropriate loudness.

I was looking forward to this. Constructing a simulator would let me dust off some woodworking equipment that had been lying dormant in the garage. It's fun to build stuff.

To simulate the inside of the scanner, we needed a tube of the right diameter. The MRI bore measured sixty centimeters, or two feet, in diameter — larger than any tubing you would find at your local hardware store. That is, however, a common diameter of concrete pillars used in the construction of buildings. These pillars are made by pouring concrete into molds sold under the trade name Sonotube. Andrew made a few calls to construction supply houses around Atlanta and soon located a twelve-foot-long, two-foot-wide Sonotube.

"Do they sell it by the foot?" I asked him.

"Nope," Andrew replied. "We have to buy the whole piece."

"How much?"

"About a hundred bucks."

"How long is the MRI bore again?"

"Six feet."

"This is great," I said. "We can do this NASA style."

Andrew looked puzzled.

"In the old days," I explained, "NASA always launched two spacecraft for their missions. The reason was that most of the cost of a project was in the design and development. Once those were achieved, the added cost of a second spacecraft was minimal. Plus, it added a level of protection if one craft failed. If we need to buy twelve feet of Sonotube, we might as well just cut it in half and build two simulators. We'll give one to Mark to use at CPT, and I can keep one at home to test on Callie."

Building the main tube didn't require much beyond cutting the Sonotube in half. Andrew and I found everything else we needed at the local Home Depot. We bought two folding tables to mount the tubes on. A sheet of plywood and some lumber would serve as the patient table inside the tube. We did the construction in my garage on a Saturday.

Even though the result looked nothing like an actual MRI scan-

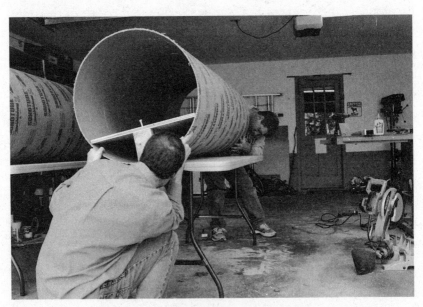

Andrew and I building the MRI simulator.
(Helen Berns)

ner on the outside, the important part for the dog was the interior of the tube. Andrew had obtained all the measurements from the real scanner. All we had to do was duplicate the height and width of the patient table when fully inserted into the MRI bore. To test it, we took turns crawling into the simulator. We both agreed it felt just as claustrophobic as the real thing.

Building a mockup of the birdcage was a little more complicated. It is a peculiar shape, a cylinder about a foot long and a foot in diameter. Because it lies on its side, the cylinder rests in a cradle that attaches to the patient table inside the MRI tube. Andrew took exact measurements of the real coil and made a full-scale tracing of the ends of the birdcage. We transferred the tracings to some plywood and used a jigsaw to cut out exact replicas of the ends of the coil. Wood dowels were used to simulate the cage of the coil and also to hold the ends of the birdcage together. We bent a thin sheet of plywood into a semicircle and glued it inside the whole construction. For humans, this would form the cradle where the head rests. The dogs would need to shimmy their whole body into the coil and assume a sphinx position.

Callie kept a respectable distance from the simulator. She wasn't afraid of it, but she wasn't treating it like a toy either.

Just for fun, I tried to coax her into the tube, but she wanted nothing to do with it. Even with a dog treat inside, she still wouldn't go in. The whole thing was just too foreign. Plus, it was elevated on a table, and she did not like being placed on a table. Too much like a visit to the vet.

Maybe the birdcage would be an easier place to start.

We began by placing the birdcage on the floor. I wanted Callie to sniff it out on her own terms. After a few minutes, she became bored and walked away. This was a good sign: she was getting used to it and didn't view it as a threat. Next, I lay down on the floor and put my head in the birdcage. Callie still wasn't ready to jump in with me. But a little peanut butter on my lips changed her mind. She bounded onto my chest and stuck her head in to lick it off.

Since she seemed to be having fun now, I dabbed a little peanut butter inside the coil to get her to go in by herself. She happily lapped it up. To avoid smearing the whole birdcage with peanut butter, I switched over to dog treats.

Each time she stuck her head in the birdcage, I moved the treat a little farther back. I wanted to see if she would assume the sphinx position in the birdcage, but I had no idea how to do that. As much as I loved Callie and secretly hoped that she was going to be subject number one, I was afraid that she was too ill behaved for the experiment.

I e-mailed Mark some pictures of the mock scanner. In the last photo, Callie lounged next to the head coil.

Much to my delight, it was Mark who suggested using her.

"She looks comfortable with it," he wrote. "Why not make Callie the first subject?"

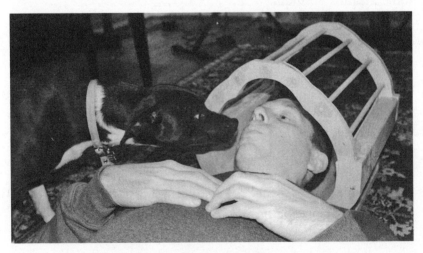

I test the mock head coil, while Callie investigates.
(Helen Berns)

9

Basic Training

CALLIE LOOKED GOOD AT HOME, but how would she do in an unfamiliar environment? She showed no fear of the head coil, a sign she would be able to adapt to novel tasks. But there was only one way to find out for sure.

Helen, eager to see how Callie would do with the training, helped me load her into the car, and the three of us headed to CPT with the head coil to see Mark work his magic.

Helen entered with Callie, while I placed the head coil on the floor.

Mark looked at it and nodded. "This should be easy. Did you bring treats?"

From puppy training, I knew that soft treats are best. You can cut them up into tiny pieces so the dog doesn't fill up too quickly. And the dog can consume them easily without getting distracted by crunching on a hard biscuit. The only treats I could find around the house were some hot dogs that had been pushed to the back of the refrigerator. I had no idea how long they'd been there, but they smelled okay, and Callie loved them. I handed Mark a baggie full of sliced-up hot dogs.

"First," he said, "let's start with the clicker."

A training clicker is a small device about the size of a USB flash drive that, unsurprisingly, makes a loud *click* when pressed. Dogs can hear the clicker from across the room. The advantage of using one is that it always makes the same sound, which is not the case with vocal commands. Because it's almost impossible to screw up, the clicker is a useful tool for beginners like me. Its operation is simple: when the dog does something correct, you click. For this to work, however, you first have to teach the dog that a click equals a reward. This is classic classical conditioning. Just like Pavlov.

Callie tracked the bag of hot dogs as I handed it to Mark. Then she dutifully sat at his feet, tail sweeping the floor. Mark clicked and immediately gave her a piece of hot dog. Callie got even more excited. She could barely sit.

At this point, what Callie was doing was unimportant. Mark periodically clicked and handed her a reward. He was establishing the association of each click with a transfer of reward, making it a conditioned stimulus. It didn't take long. A dozen click-rewards, and Callie understood the association. With the meaning of the clicker established, Callie was ready to learn a behavior. I could immediately see how the clicker was going to make this easier.

Mark explained another advantage of using the clicker. "We are going to shape her behavior. Initially, anything Callie does that is close to the desired behavior will be rewarded. The clicker makes it absolutely clear to her that she has done something correctly. This way, she won't get conditioned to just my voice or your voice."

The clicker gives instantaneous feedback, making it clear to a dog that she has done something good without wasting time fumbling for the treats. Unlike a human, a dog's memory for what she has just done appears to be very limited. The longer the interval between the desired behavior and the subsequent reward, the less likely the dog will make the association. This phenomenon is called *temporal discounting*. Research in rats suggests that a reward given four seconds after a desired behavior is roughly half as effective as one given immediately. If the handler is deeply involved with the dog, using hand signals and vocal commands, he might not be able to give a reward for a while. This is especially true of complex behaviors. The clicker solves this problem by giving instantaneous feedback.

Mark was beginning to lure Callie into the head coil. Reaching into the coil with a hot dog in one hand and the clicker in the other, Mark had already succeeded in getting Callie to place her nose inside. Each time she did so, Mark clicked, praised her, and gave her a bit of hot dog.

With every click-reward, Mark pulled the food back a little bit, shaping her behavior gradually. Within ten repetitions, he had Callie crouching in the coil with her snout poking out the other end. Some gentle pressure on her rump indicated that she should lie down in the coil. As soon as she did, Mark clicked and exclaimed, "Good coil!" Callie wagged her tail and licked the hot dog from his hand.

I couldn't believe how quickly Mark had gotten Callie where she needed to be.

"How is the positioning?" he asked.

Callie was lying down in a sphinx position in the coil. Her paws hung over the near edge. She would need to move back a little bit.

"We'll want her head in the center." Mark nudged her back an inch and clicked.

"You can shape her behavior at home too," he said. "I think she'll do really well with this."

A woman walked into CPT with a border collie.

"This is Melissa Cate," Mark said. "Melissa runs some of our agility classes at CPT. She's interested in volunteering her dog for the MRI."

"Mark told me about the Dog Project." Pointing to her dog, she said, "This is McKenzie."

McKenzie was Melissa's three-year-old border collie. Melissa had begun agility competitions a few years earlier with her boxer, Zeke, who had reached the highest ranks. Zeke was now eight years old and slowing down a bit, so Melissa had gotten McKenzie as a puppy to keep competing in agility. They had been going strong ever since.

McKenzie was leggy and lean, about thirty-five pounds, with a long, thin head that would easily fit in the head coil. She trotted over to me and stared long and hard. She quickly realized that I was not a herdable animal and moved on to check out Helen.

Callie zoomed over and assumed a play bow with her front legs flat and her rump in the air, tail wagging like a vibrating string. We let the two dogs off-leash and they ran around the room. Callie did orbits around McKenzie, who seemed indifferent to the newbie dog.

It was time for McKenzie's try with the head coil. With a dog treat, Melissa had no trouble coaxing her into the coil. Nibbling the food out of Melissa's hand, McKenzie appeared unaware of the coil altogether. In agility competition, the dogs run through a serpentine tunnel, and McKenzie was completely comfortable in an enclosed space.

After a few minutes, Melissa commanded McKenzie to lie down. "*Platz*," she said, using the German word for "down." Mark explained that German words are commonly used in dog training because of the popular Schutzhund competitions. These began as training programs and tests for German shepherds but evolved into a full-fledged sport involving tracking, obedience, and protection phases.

With McKenzie lying down in the head coil, Melissa backed away to the other side of the room. McKenzie didn't budge. In fact, she stayed motionless for a solid minute. When I saw what a well-trained dog like McKenzie could do, I knew we could really do this. If the dogs would go into the head coil, they would go into the MRI.

• • •

Melissa working with McKenzie in the head coil.
Callie watches from across the room.
(Bryan Meltz)

So far, Mark and Melissa had been using basic behaviorist techniques. The appeal of behaviorism in dog training is its simplicity. By making rewards like food and praise contingent on desired behaviors, dogs quickly learn what to do to get something they want. But what do dogs think of this? After all, they aren't robots bumbling around the world, randomly doing things and finding out which behaviors result in food. Dogs show purposeful and consistent behavior whether humans are there or not. This suggests that dogs have some internal mental model of how things work in their world. It is a limited model, of course. For instance, they don't understand technology like computers or television. But dogs do understand how to get along with each other and with other species like humans, which is not an insignificant skill, and they don't need treats to learn how.

While McKenzie remained still in the head coil, Callie watched with rapt attention. It's possible she was interested in only the food being handed out, but her gaze wasn't always tracking Melissa's hands. Callie's eyes darted back and forth between McKenzie and Melissa. You could almost see the wheels turning inside Callie's head as she tried to figure out what was going on.

Even though we were using basic principles of behaviorism like positive reinforcement and shaping to train the dogs to enter and lie still in the head coil, Callie made it clear that a different type of learning was going on too. She was learning by observation.

Social learning, or imitation, is an obvious feature of human behavior. We can learn a great deal simply by observing what other people do. Strangely, dogs have not been given much credit for being able to do this too. But Callie illustrated clearly what everyone with more than one dog knows: dogs learn from each other.

Although behaviorist experiments dominate the canine research literature, there have been a few experiments demonstrating social learning between dogs. An old experiment found that puppies that observe littermates pulling a cart by a string can copy that behavior. Another study showed that puppies that watched their mother, a police dog, search for narcotics did better when learning this task compared to puppies that did not observe their mother first.

Relatively little is known about the neurobiology of social learning or imitative behavior. Even in humans, we don't know much about which parts of the brain are involved. Whether dog or human, social learning is not contingent on rewards. So why does the puppy copy its littermates and pull the cart? After all, there is no food to be gained. Maybe the Dog Project would provide the answers.

But first, we needed to train Callie and McKenzie to perform a complicated task under noisy conditions.

Over the next few weeks, Callie and I worked with the head coil at home on a daily basis. I limited the sessions to ten minutes. Mark had explained that short, daily training is much more effective than infrequent long sessions. This prevents both the dog and the handler from getting bored. Consistency is the key.

Callie learned quickly. As soon as I reached for the head coil, Callie would jump up and down to try to get in it. Once inside, she assumed the sphinx position and waited for me to give her hot dogs. Her tail never stopped wagging.

"Good coil!" I would praise. More tail wags.

The next step was to introduce the chin rest. When we scan humans, the subject normally lies on his back with his head surrounded by foam padding in the head coil. The padding makes it comfortable while also preventing head movement. But the human setup wouldn't work for a dog. Callie and McKenzie would have to lay on their stomachs, and I doubted that either of them would want her head surrounded by foam, like Leonard did with his monkeys at Yerkes.

I didn't yet know how we were going to solve the head movement problem. The first step, though, was to give the dogs something they could rest their heads on. Something firm, yet comfortable. My first thought was the foam used in seat cushions, so I picked up the firmest foam I could find at the fabric store, cut off a chunk, and, while Callie was relaxing on the sofa, gently wedged it under her chin. She just let her head sink into it and went to sleep. That was a good sign, but the foam compressed too much to offer enough support. We needed something firmer.

I went to the hardware store and looked at foam insulation. Too hard. I was beginning to feel like Goldilocks. For several days, I searched in vain for something that would work. The chin rest would have to span the diameter of the foot-wide head coil. At that length, furniture foam collapsed with the slightest pressure in the middle of the span. But the materials at the hardware store were uncomfortably stiff.

The solution caught my eye in a sporting goods store. Helen and Maddy had wanted some new sneakers. While they were trying them on, I milled about, not looking for anything in particular. It was the middle of December, and the store was having a sale on summer swim gear. A stack of boogie boards was pushed into a corner with a handwritten sign: $5 each. I squeezed one. Firm, but not hard.

I took the whole stack to the register. The cashier looked at me like I was crazy.

"Science project," I said.

At home, I used a utility knife to cut a strip of boogie board to match the inner diameter of the head coil. It made a comfortable bridge that didn't collapse in the middle under pressure.

I got my baggie of hot dogs and approached Callie with the foam bar. She saw the treats and started wagging her tail. With her in the sphinx position on the floor, I gently pushed the foam under her chin.

"Touch," I said, and gave her a treat.

As usual, Callie was a quick study. After a few repetitions, her head relaxed as soon as I gave the "touch" command, and I could feel the weight of her head against the foam bar. Pretty soon, I didn't even have to cue her by pressing the boogie board against her chin. With the bar a centimeter below her chin, she would drop her head to make contact on command.

The next day, we practiced the "touch" command with Callie in the head coil. She got it. With the foam rest spanning the diameter,

Callie scooted in and stuck her paws beneath it. I said, "Touch," and she plunked her head on the bar.

"Good girl!" I exclaimed. She just wagged her tail. I couldn't believe how quickly she was picking this stuff up. Lyra, drooling nearby, soon learned that she too would get hot dogs by hanging around the head coil. Lyra would then start barking if she didn't get a piece of the action. She had a sensitive stomach, though, and tended to burp if I gave her too many hot dogs.

Each day we practiced with the chin rest in the head coil, and each day Callie held her head in position for longer and longer stretches. After a week of daily sessions, I no longer needed to say the command. I would just put the head coil on the floor, and she would go right in and plop her head on the rest.

We were gradually making the task more complex for Callie and McKenzie by adding elements before the final behavior. The technique is called *backward chaining*. Once Callie had learned to go into the coil, I placed the coil inside the mockup of the MRI bore. Since Callie knew that she would be rewarded only for going into the head coil, she trotted into the tube and into the coil, after which I promptly rewarded her. Next, I raised the tube to the height of the patient table of a real MRI machine. Callie would have to be taught to go up a set of doggie steps. These were designed for dogs to walk up to the height of their owners' beds. Since they were made entirely out of plastic, they would be safe for use next to the real MRI.

It took a couple of days to teach Callie to go up the steps. I started by placing a hot dog on each step. Callie followed the trail of meat right to the top, where I gave her excessive praise. Once she was used to the steps, I placed them in front of the elevated tube and continued the meat trail all the way to the head coil. Once she was in the tube, I ran around to the other end and pointed to the head coil. She scooted in and waited for more treats.

The last and most challenging element was the scanner noise. Andrew had already recorded the jackhammer-like sounds of the MRI in action. Initially, we focused just on getting Callie and McKenzie used to the level of ambient noise. Later, we would need to figure out the exact scanner settings for the dogs, which would result in slightly different sounds.

I started by simply playing the scanner noise at low volume through a stereo while I did the training with Callie. She quickly learned to

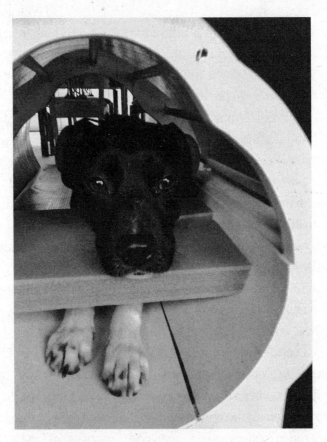

Callie sporting the earmuffs and learning to use
the boogie board chin rest in the head coil.
(Gregory Berns)

ignore it. Each day I would increase the volume a little bit. Soon, though, it would get to a level that was unpleasant. Time to introduce Callie to the earmuffs.

Human subjects wear earplugs, but I had yet to meet a dog that would let you put anything in his ears. Plus, the ear canal of a dog makes a right-angle bend. If an earplug became wedged in the canal beyond the bend, we might not be able to get it out. The only alternative was earmuffs that went over the outside of each ear. Amazingly, I discovered Safe and Sound Pets, a company that makes Mutt Muffs. The founder of the company, a pilot of small aircraft, realized the need to protect his dog's hearing when taking him up for flights. He adapted human earmuffs to a more triangular shape that would fit most dogs' heads. We ordered several sets in different sizes.

Callie was not so thrilled with the Mutt Muffs. It took many bits of hot dog, first for putting her nose through the loop formed by the chin strap and eventually for letting me push the earmuffs back over her head. Even then, she would paw them off right away. I didn't push her. With Mark's advice, I gradually lengthened the time she had to wear them before giving her a piece of hot dog. Pretty soon, Callie would leave the earmuffs on, trot up the steps into the tube, and place her head on the chin rest in the head coil.

Everything was going faster than I had anticipated. Melissa was working with McKenzie in parallel; unsurprisingly, they made even quicker progress. In fact, the training was going so well, I hadn't even stopped to ponder what we were doing to their brains, which, after all, was the subject of the Dog Project.

Even though we were using basic behaviorist principles to shape a complex behavior in the dogs, it couldn't explain what the dogs thought of all this. If we cared about just behavior, the reason a dog did something or what she thought wouldn't matter. But if we got to the point of actually scanning their brains, the dogs' motivations could have a big impact on what we found. Doing something for food would look very different from doing something for social praise or, dare I say, love.

I had the nagging feeling that the ease with which dogs slip into human lives could not be fully explained by behaviorist theories. For dogs to do what they do, they must have a rich inner life that goes beyond a chain of actions resulting in food. Dogs must have a rich mental model of their environment. As highly social animals, these mental models are likely to be weighted heavily toward social relationships. Not just dominance and subordination, but more fluid models of how they should behave with members of their household, either dog or human, and how these interactions will affect their current state of well-being.

It makes you wonder who is training whom. Skinner and Pavlov were partly right. Their principles are highly effective in training behavior. But they studied animals in a laboratory — a place where the human controls pleasure and pain. Dogs in their native environment — the human household — interact with us in a much more natural way. There is give-and-take, and testing, on both sides.

10

The Stand-In

W ITH CALLIE AND MCKENZIE making rapid progress in their training, we would soon be ready to make
the jump to the real scanner. Although the mock head
coil and tube were good facsimiles of the MRI, they weren't the real
things. There was no way to simulate the smells and sounds of the
hospital, for instance. We wouldn't know how the dogs would react
until we actually brought them there.

The initial introduction to the real MRI would be critical. Dogs
can form negative imprints of environments based on one event: a
loud noise such as a slamming door, an encounter with someone who
doesn't like dogs. Any of these could permanently affect a dog's impression of the MRI. If that happened, and the dog didn't want to go
near the scanner as a result, all the training we'd done would have
been wasted.

Mark and I were particularly concerned about noise. The MRI
scanner makes a wide range of sounds. The magnetic field is always
on, and this requires constant tending by an array of devices, like the
pumps that circulate chilled water around the magnet. When you
enter the room, the first thing you hear is the heartbeat of the circu-

lation pumps. If you listen carefully, you will also hear the machine breathing, the sound of the "cold-head"—a compressor that keeps the helium under pressure.

How would the dogs react to this living machine?

When scanning a subject, MRIs are loud. Depending on the particular settings, an MRI can reach nearly 100 decibels. Every 6 decibels means a doubling in the sound pressure. Normal conversation is about 60 decibels. Busy traffic, 80 or 90. A jackhammer is 100. Hearing damage for humans begins at around 120 decibels, equivalent to a jet engine at a hundred meters. Nobody knows at what level hearing damage occurs in dogs, but one point of reference is hunting. The report of a hunting-caliber bullet is 170 decibels, and hearing loss after repeated gunshot exposure is a well-known phenomenon in both hunting and military working dogs.

Earmuffs can cut sound levels by 20 or even 30 decibels, so assuming that a dog's hearing is more sensitive than that of a human, Callie and McKenzie would be fine as long as they wore ear protection.

The loudness of the MRI wasn't the whole story, though. The type of sound made could also have a big effect on the dogs. The MRI makes different types of sounds depending on the type of brain scan being performed. Some scans sound like a swarm of bees, while other scans are like the klaxon of a submarine preparing to dive. The specific sound depends on dozens of parameters that are programmed for each scan. These parameters indicate how many slices through the brain will be made, how thick they will be, and whether they should focus on gray matter (the neurons) or white matter (the connections between neurons) or on changes in blood flow like we do in fMRI.

The only way to get an accurate recording of the sounds the dogs would experience during their scans would be to program the actual scan sequence with the exact parameters necessary to scan the dogs' brains. But since nobody had scanned a dog's brain before, at least not with fMRI, we had no idea what the correct settings might be.

With a dog in the scanner, we could figure out the correct settings, but we needed the right settings in order to record the sounds to train him to get him into the scanner in the first place.

I felt like a dog chasing his tail.

At lab meeting, I brought up the conundrum.

"What if you use the standard human settings to record the scanner noise?" Lisa suggested.

"It might be good enough," I said. "But what if it isn't?"

"I bet the dogs could tell the difference," Andrew said. "If we train them with the wrong sounds, they might freak out when they hear the real thing."

"We need a stand-in," I said. "Something that can take the place of the dog while we fiddle with the scanner settings."

Lisa's forehead knitted up in thought. Everyone else looked at the floor. Before anyone suggested it, I headed off the obvious.

"We're not using a dead dog."

"Why don't you just go to the supermarket and buy a steak and scan that?" Gavin joked.

"You mean like the famous dead-salmon study?" Andrew asked.

A few years before, neuroscientists had used fMRI to scan a salmon purchased at a local fish market. As they wrote in their findings, the fish "was not alive at the time of scanning." They presented their results at a conference, but most scientists dismissed it as a joke. It wasn't. The point was to measure the accuracy of fMRI and how the technique could sometimes lead to the appearance of brain activations that weren't actually there. Obviously a dead salmon couldn't have brain activation, but the scientists showed that with poor statistical technique, it might appear that way.

Gavin's joke wasn't half bad. But a steak (or a salmon) would be a lousy stand-in for a dog.

"We need something more doglike," I said.

"A pig?" Gavin said.

"Too big."

"How about a lamb?" Andrew suggested.

"Can you buy a whole lamb at the market?" I wondered.

After a few phone calls to some local butchers, Andrew found a lead. It wasn't a whole lamb — you needed to get that directly from a farm — but there was a market that might sell us a lamb's head.

"I think he said they get their delivery of lamb heads on Wednesdays," Andrew explained. "I'm not completely sure because I couldn't understand some of what he was saying. But he definitely said they go fast."

"Today is Wednesday," I pointed out.

"Giddyup!"

The halal meat market had no sign. The "market" consisted of a counter at the rear of a convenience store, itself sandwiched in a run-down strip mall and sharing a wall with a video store specializing in bootlegged Middle Eastern movies.

Andrew and I walked in to find a trio of bearded men hanging out at the cash register, smoking cigarettes and watching soccer on TV. They said nothing as we made our way to the rear of the store. I noticed some elaborate water pipes on display.

At the butcher's counter, a spread of organ meats glistened beneath the glass case. Kidneys I recognized. The rest — not so much. The animals of origin were a mystery to me too.

A squat guy wearing a tight soccer jersey peered over the counter.

"You guy call about lamb head?" he asked in a Middle Eastern accent.

"Yes."

"How many you want?"

Andrew and I looked at each other.

"How many do you have?" I replied.

"Lots."

We conferred briefly and decided that we should have a backup in case something went wrong.

"Two," I said.

The butcher disappeared through a doorway covered with vinyl slats. A moment later he returned and deposited two heads on the counter with an authoritative clank.

"They're frozen," I said.

"Yes," said the butcher, "fresh frozen."

They bore a resemblance to a lamb, but as all the wool had been removed, it was hard to tell what they were. The lips had retracted a bit, and the faces were fixed in permanent grimaces.

The size was right, I had to admit. In fact, they were about the same size as Lyra's head. I shivered and pushed that unpleasant image out of my mind.

"Where is the rest of the lamb?" I asked.

"Just head," he replied.

"Do they still have their brains?"

The butcher brought his fingers to his mouth in the sign known to foodies around the world and said, "Yes. Delicacy."

Ideally, we would have gotten a whole lamb to stand in for a dog. Anything you put inside an MRI disturbs the magnetic field. The bigger the object, the greater the disturbance, and as the scanner compensates for these disturbances, it makes different kinds of sounds. The lamb's head was not going to have enough mass to replicate the disturbance created by a dog. We needed something else.

Andrew pointed to a pair of hooves in the butcher's case. They appeared to be the front legs of a calf starting just above the ankle joint.

In the actual MRI, the dogs would be scanned in a sphinx position. Their heads would be upright, supported by a chin rest, and their front paws would be sticking straight forward. Andrew realized we could use the calf hooves to simulate the front paws of the dog.

The combination would give a close approximation to the shape and mass of the part of the dog that would be at the center of the scanner. We paid for our meats and headed back to the lab with two lamb heads and a pair of calf hooves.

The vegetarians in the lab weren't going to be happy.

We let the heads thaw overnight and reserved time on the MRI scanner for the following evening. Scanning dead animal parts in the MRI is the kind of thing best done discreetly. Once thawed, the heads, now swimming in their own juices, looked even worse. Their eyes had taken on an opaque haze. Andrew and I double-bagged everything and headed to the scanner.

We were greeted by Lei Zhou, a Chinese postdoc on duty that evening. Lei had received his PhD in physics and was intimately familiar with the technical wizardry behind MRI. His English, however, had

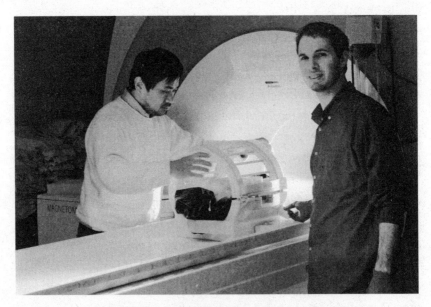

Lei and Andrew preparing to scan the lamb's head.
(*Gregory Berns*)

a ways to go. I could only hope that we understood each other during this unusual procedure.

Andrew unloaded our cargo, and we proceeded to arrange it in the head coil of the scanner. With foam pads propping up the body parts, Lei snapped on the top of the coil and sent the whole mess into the center of the scanner.

When you place something in the MRI, the magnetic field tugs on the atoms inside the object. In living tissue or, as in the case of the lamb's head, formerly living tissue, hydrogen is the most common atom. There are two hydrogen atoms in every water molecule, and water accounts for 60 percent of body weight in humans. Hydrogen is also abundant in the brain. The outer membranes of neurons and their supporting cells, called glia, are rich in fat and cholesterol, which have large numbers of hydrogen atoms.

A hydrogen atom has one proton and one electron. The proton is like a spinning top. Normally, the protons spin in random directions, but inside the MRI they line up with the magnetic field. Like spinning tops, the protons also wobble a little bit. The stronger the magnetic field, the faster they wobble. If you hit the protons with radio waves exactly in sync with their wobbling, the protons jump into a higher energy state. This is called *magnetic resonance*. Different types of atoms resonate at different frequencies. For the strength of scanner we use, hydrogen resonates at 127 MHz, which falls in the range of radio waves — just beyond the FM dial. Carbon, another common element in the body, resonates at 32 MHz. MRI works by sending in a blast of radio waves that excite the atom of interest — in most cases hydrogen because of its abundance and superior sensitivity to magnetic fields. When the radio waves are turned off, the protons relax back to their original state and, in the process, cause an oscillating magnetic field that can be picked up by an antenna. The head coil is nothing more than a fancy FM radio antenna that picks up these signals from the protons in the brain.

Not all protons behave the same way. The protons in a water mol-ecule wobble slightly differently than the protons in a fat molecule. These slight differences can be detected by the MRI and, with the help of a computer, be used to construct a visual image representing the types and locations of these different molecules.

We would need to do three types of scans on each subject. The localizer, which lasts only a few seconds, gives a snapshot of the loca-tion and orientation of the head in the magnet. The localizer scan of the lamb's head came out well. We could clearly make out the brain. The human settings for the localizer seemed to work. Next up was the structural image. For humans, we like as much anatomical detail as possible, but this has to be weighed against the time it takes to get high-resolution images. Images clear enough to resolve features as small as one millimeter take six minutes to complete. Humans have no problem holding still for that long, but there was no way our dogs would. I told Lei that we needed to come up with a structural se-quence that would take no more than thirty seconds. I figured that would be the limit for most dogs.

This turned out to be somewhat difficult. The normal structural scans couldn't be completed that quickly, so we had to switch to a different type of scan. This one didn't show as much detail, but we were able to find a combination of parameters that produced a usable image in under thirty seconds.

We spent an awful lot of time figuring out the best orientation of the brain. If you think of the MRI as being a digital bread slicer, we had to decide which way to cut the slices: left to right, top to bottom, or front to back. Since the human head is pretty close to a sphere, it doesn't make a whole lot of difference which way you slice it. But a dog's head, like the lamb's, is elongated front to back and generally pretty flat from top to bottom.

As the images of the lamb's head came up on the screen, we saw how little of the head was actually occupied by brain. Most of it was

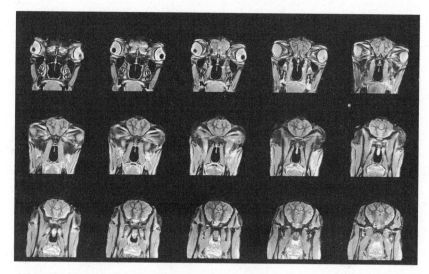

Anatomical images of the lamb's head. The slices go from front to back. The eyeballs
are visible in the top row, while the brain appears prominently in the middle and
lower rows. The large black cavities are nasal sinuses.
(Gregory Berns)

nose and muscle. Those air pockets in the nose can wreak havoc
with the MRI images too. Abrupt transitions in tissue density, such
as going from air to skull, cause distortions in the magnetic field,
which result in warped images. By carefully selecting the orientation
of the slices, you can minimize this effect. Slicing from front to back
seemed to give us the best results.

Finally, it was time to attempt some functional scans, which are
two-second glimpses of the brain in action. By continuously acquiring these functional scans while the subject does something, we can
measure changes in brain activity. Think of the functional scans as
the individual frames of a movie. Even though each one takes only
two seconds, the subject might be in the scanner for half an hour
during functional scanning. During such a session, we would acquire
nine hundred functional images, at a rate of thirty scans a minute for
thirty minutes.

Of course, the lamb was dead, so we didn't expect to see much "activation." But we only needed to figure out how many slices it took to cover the brain and how to orient the brain for the most efficient coverage. Once we worked that out, Andrew and I recorded the sounds of the scanner running this sequence.

We could now introduce Callie and McKenzie to the actual noise they would experience in the scanner and gradually let them get used to it.

11

The Carrot or the Stick?

THE CHALLENGE OF ENTERING the head coil and placing her chin on the boogie board chin rest had long been overcome. As soon as Callie heard the rustling of the plastic baggie containing bits of chopped-up hot dog, she knew. She would bound into the kitchen, wagging her whole rear end, and look at me with excitement and anticipation.

"Wanna do some training?" I would ask in a high-pitched voice.

Our training regimen had outgrown the basement. The only room in the house big enough to contain what was now a full-blown MRI simulator was the living room. Kat eyed the monstrosity in her living room, a space formerly occupied by an elegant sofa set and coffee table now pushed off to the side.

"There isn't any other place for this?" she asked.

"It's too heavy to move down in the basement," I replied. "And I don't think it will fit through the door."

"You mean you constructed this in the living room without a way to get it out?"

"No, no," I reassured her. "It comes apart."

I had dusted off a PA system left over from my guitar-swinging days in a garage band. As I set the speakers on a stand facing the tube, Helen came into the living room.

"What's that for?"

"To simulate the noise of the scanner," I explained. "It's the only thing we have that's loud enough."

She nodded, and together we snaked cables from the speakers to the amplifier. We aimed one speaker at the side of the tube to simulate the vibrations that course through the MRI. The other speaker went at the end of the tube to achieve the full decibel level inside.

"Daddy?"

"Yes?"

"Can I come with you when you scan Callie?"

This question took me by surprise. I wondered what had motivated it.

"Why do you want to see the experiment?" I asked. "It might not even work."

"I know, but I want to see it," she said.

"Is it because you want to skip school?"

She turned away and mumbled, "Maybe." But quickly recovering, she continued. "That's not the main reason. I really want to see the experiment. Don't you always say that real science is exciting? Wouldn't I learn more there than I would at school?"

In that, her logic was flawless.

"I'll have to think about it."

I had no doubt that Helen would learn more about science watching this experiment than she would in an entire week of science class at the middle school.

This was Helen's first year of middle school, and the transition from elementary school had been a shock for all of us. Her workload was so much larger than what she was used to, she still hadn't quite figured out how to balance homework and fun. In addition to the usual math, English, and social studies, her school required Latin for all sixth graders. In a classic case of confusing correlation with causation, the curriculum committee cited studies showing that kids who learned Latin did better on the SAT and reasoned that if all kids took Latin, their test scores would improve. Unfortunately, just because kids who take Latin have higher SAT scores doesn't mean that Latin is the cause. These kids might already have a larger vocabulary and an interest in learning another language.

Latin wasn't the problem, though. Much to my dismay, it was science.

Early in the year, I had tried to explain to Helen that science is always changing.

To which she asked, "You mean that this stuff is wrong?"

"Some of it."

"Then why am I learning it?"

Because the state says you have to, I thought. But what I said was "Science is a way of answering questions about the world around us. What you are learning is our current understanding of the universe. As we learn more, our understanding changes."

"I still hate it."

I understood her frustration. She really had tried to memorize facts of geology and the weather systems of the Mississippi Delta and Piedmont regions of the Southeast. But no matter how hard she studied, the science teacher seemed to throw obscure questions at the kids that I would have thought more appropriate for a high school or introductory college class.

The new semester had just begun when Helen asked me about coming along for the first scans. The next morning, we were standing at the bus stop.

"Do you have any tests this week?" I asked. Her face turned white.

"What day is it?"

"Tuesday."

"I think I have a science test today."

I was furious. "Helen, you just had a three-day weekend, and you didn't study at all?"

"I forgot."

"And, on top of that, you want to skip school to see the Dog Project?"

Her eyes were starting to tear up. "I know the material."

"How can you know the material if you didn't study?"

There was no answer to the question. She got on the bus, and I walked home frustrated that she wasn't prioritizing her schoolwork.

Both Kat and I were ready to ground Helen as soon as she got home that night. In the past, when she received less than an 80 on her tests, she would lose computer privileges until she brought her grade up. While this had been an effective strategy to prevent her from wasting time on computer games, the time that was freed up didn't generally translate into more studying. Computer time was exchanged for sulking time. She was well on her way to perfecting the art of the silent treatment.

From the day she joined our household, it was clear that Callie was an alpha dog. In Cesar Millan's terminology, she wanted to be pack leader. She hopped on furniture whenever she wanted. When Lyra started to chew on a bone, Callie would dart in to take it away, only to drop it on the ground a few feet away, indicating that she determined who would be allowed to eat the "prey." And when Callie settled into our bed at night, it was almost always in a position uncomfortable for the human occupants. The expression "let sleeping dogs lie" should have been broadcast above her head, because any attempts to move her to a more harmonious location were met with the most vicious snarling possible from the little creature.

Similarly, her insatiable appetite meant that all food had to be pushed back from the edge of the kitchen counter. With her long snout, she could grasp any morsel of food within three inches, even if she couldn't see it. She once licked clean precisely half of a pumpkin pie, which was the extent her tongue could reach. Every time we caught her with paws up on the counter, we yelled at her to get down. Although she complied with the command, it never prevented her from doing it again, usually within minutes.

It was this behavioral stubbornness that made us doubt Callie's ability to participate in the Dog Project. I eventually realized that the problematic aspects of her behavior had nothing to do with her ability to learn. She could not only learn a complex task like going into the MRI scanner, she could actually learn to enjoy it.

I wondered whether something similar was going on with Helen.

Every time she did poorly on a science test, we used the equivalent of a squirt bottle or a shake can to curb the behavior: a scolding followed by a mild punishment. Punishment can be very effective in shaping behavior, but it works only when there is a credible threat of punishment present. This is so important, it bears repeating: only the *threat* of punishment can change behavior. As soon as the threat disappears, behavior reverts to its natural state. Punishment after the fact serves only to establish a credible threat in the future but does nothing to change what has already happened.

Helen's lack of studying was water under the bridge. Grounding her would not change the inevitable poor grade she was about to receive. Would it make her study more in the future? Possibly, but only under the constant threat of punishment. There had to be a better way.

I asked Kat what she thought.

"I don't like the idea of punishing our kids for not studying either," she said.

"I wish Helen would want to study," I said. "But if I had to study from that textbook, I probably wouldn't do it either."

"What can we do?" Kat asked.

"Maybe we need more of the carrot and less of the stick."

We put our plan into motion at the dinner table that night. Not surprisingly, Helen didn't think she had done very well on the test and picked at her food sullenly. Maddy sensed the tension and remained silent.

With great solemnity, I announced, "Mommy and I have been thinking very seriously about the Dog Project."

Bracing for the inevitable hammer about to fall, Helen didn't look up from her plate.

"Helen," I continued, "you really want to see the scanning on the big day?"

"Yes," she pleaded.

"Okay. Mommy and I have discussed this, and because this is so special and may never happen again, we want to let you go."

"Really?" she exclaimed.

"This is important to you?" I asked.

She nodded vigorously.

"Good," I continued, "because there is a condition."

"What?" Helen asked.

"In order for you to miss school, you have to pull your science grade up to an A," I explained. "If you have an A in the class, you can come. The Dog Project is very important to me, and I would really like you to be there to share in it."

"I can do that!" she agreed.

For the next several days, the prospect of positive reinforcement had the desired effect. Although Helen still didn't enjoy studying science, there was a noticeable decrease in homework resistance. She threw herself into making flash cards and made an earnest attempt to memorize the material. Kat and I patted ourselves on our backs, celebrating our success at applying dog-training theory to preteen behaviorism.

But like dog training, the effectiveness is in the details.

Callie was making progress in the training in large part because I was beginning to learn how to make it clear what I expected of her. Baby steps, coupled with consistent reward, make for effective learning. But if the desired behavior is too difficult, then the reward becomes unobtainable and motivation declines.

With Helen, the desired behavior was clear: get an A. But what I had neglected to consider was the inherent unpredictability of her science teacher. I mistakenly assumed that if Helen put in the necessary effort, she would be rewarded with a good grade.

Big mistake.

A week later, despite all of her efforts, she returned home with a 75 on a quiz. This pretty much put out of reach the possibility of raising her grade to an A, at least by the time the Dog Project launched.

"I really tried," she said. "He makes the tests too hard."

Now Kat and I were in a difficult position. Helen had failed to achieve the goal we had set. If this were Callie, I would simply make her try again until she did what I wanted. But not only were we running out of time with the scan day a week away, but I also hadn't accounted for an element out of my control: the fairness of the test.

Certainly Helen could have tried harder. With half the school year gone by, she knew what the tests were like. But that wasn't really the point. She had done what I had asked, which was to redouble her efforts at studying.

The great compromise that emerged from this hand-wringing was an explicit and concrete statement of what was expected, a goal that was entirely within her control.

"I still want you to see the dog scanning," I said. "I know the tests are picky. So how about you put in an extra hour of studying each day until the scanning?"

"If I do that, I can come?"

"Yes, but to make sure that you're studying the right things, it has to be with either me or Mommy."

Helen already had one to two hours of homework each day, so this was not greeted with enthusiasm. But grudging acceptance was all that was required.

She refused to study with me that night. But over the next two days, the resentment diminished, and Helen allowed me into her room to go over concepts from science and math. I hoped that my explanations of how things worked would somehow help her remember the laundry list of facts that she would be tested on. But all I really wanted was an excuse for her to share in the excitement of the Dog Project and see what real science looked like.

12

Dogs at Work

THE DRESS REHEARSAL with Callie at the scanner made it clear that the dogs should be conditioned to more than the MRI. They needed to get used to the entire experience. We wanted them peaceful and poised on the day of scanning. The more we could do to get them used to the environment, the calmer they would eventually be. Because of her agility competitions, McKenzie was a certified road warrior, and traveling didn't faze her. But Callie was a homebody, and she didn't take well to car travel. After all, most of her car trips ended at the vet for a series of shots or a similar indignity.

So I started bringing Callie to work.

Getting her into the car was the hardest part. I would say, "Wanna go to work?" and Callie would run over to the garage door and leap up and down as though her legs were made of springs. But once I opened the car door, she would balk, tail between her legs. She would stiffen up as I placed her in the front seat. Even when we got moving, she never relaxed and would try to sit in my lap as I drove. Eventually we settled into a mutually acceptable position with her in a standing position, hind legs on the passenger seat and front legs on the center console, facing me. She shivered for the entire thirty-minute trip from house to campus. Her nervousness also caused her to shed, leaving short black hairs all over the seats.

Once we got to Emory, Callie became her normal, cheerful self. The short walk from the parking deck to the lab triggered smiles in all whom we passed. Callie liked to hop up on a stone wall, about waist high, in front of the lab building, where she would trot along, doing her best imitation of a circus dog on a tightrope.

Inside the lab, she would zoom around to each of the waste cans, looking for food scraps. Once she was satisfied there was no free food, she would interrogate the people. Lisa would lower her face to dog level and coo, "Callie!" Callie would stand on her hind legs to lick Lisa's face. The guys were friendly, if not as demonstrative, and tried to engage Callie by throwing a tennis ball around. But Callie was not a retriever. Her interest in things that moved tended toward small, furry animals.

With each trip to the lab, I brought a toy to keep her amused. It wasn't long before bones and Kongs lay scattered on the floor. A water bowl was in one corner, a doggie bed in another. The lab was starting to feel like home.

Presciently, we had included language in the official IACUC protocol specifying that the dogs would first be familiarized with the scanner environment. This would minimize the chance of the dogs freaking out and running amok. Although the intent was to placate the risk-averse lawyers, there was now the obvious side benefit that the dogs would not only have to be familiarized with the scanner, but they would need to be familiarized with the staging area—the lab. Therefore, when I brought Callie to work, I was just following protocol.

Also according to our protocol, we would need to find the right subjects. Mark had suggested a laundry list of ideal characteristics: calm, good in novel environments, good with strangers, good with other dogs, inquisitive, unafraid of loud noises, unafraid of heights, and able to wear earmuffs. These traits were specified in the official IACUC protocol that gave us permission to do the research.

Never mind that Callie and McKenzie had already been selected as our first two subjects. We would still need more dogs. We needed backups in case Callie or McKenzie couldn't make it into the MRI. Of course, we could conduct dog tryouts at CPT, and eventually we would, but we could just as easily hold "auditions" at the lab. Because the dogs hadn't yet qualified to be research subjects, and therefore fall under the IACUC rules, they existed in a gray zone between pet and research, and, as I was painfully aware, pets were not allowed.

One day, Andrew brought in his toy poodle, Daisy. Andrew had warned us that she was a temperamental dog and barked when anxious, which was often. We were already testing the boundaries of research rules, but if we got noise complaints, dogs would not be welcome anymore. Daisy was on good behavior, though. She didn't stray far from Andrew, and he limited the duration of her visit. He didn't dare bring his other dog, an American Eskimo named Mochi. She tended to leave puddles wherever she got excited. Other lab members soon followed suit. One day I was greeted by two beautiful huskies, London and Reyna. Another day, Lisa's goldendoodle, Sheriff, paid a visit. Sheriff was a golden, frizzy cross between a golden retriever and a standard poodle. He didn't qualify for the Dog Project based on size alone.

The dogs had a noticeable effect on morale. The lab felt more relaxed. The students were less distressed by whatever problems were cropping up in their research. The simple brush of a dog walking by, or the press of a cold, wet nose on your hand, was enough to drop anyone's stress level. People laughed more.

The beneficial effects of dogs in the workplace have been well documented. Sandra Barker, a professor at Virginia Commonwealth

University and director of the Center for Human-Animal Interaction, has been studying the effects of pets at work for more than a decade. In 2012, her team measured stress levels of workers who were allowed to bring their dogs to work. Normally, stress is lowest in the morning and rises steadily throughout the day. But the presence of dogs kept self-reported stress at their morning levels all day long. The researchers also found that the presence of dogs increased communication between workers.

Whether these effects on stress are simply a matter of perception has been difficult to determine. The most concrete proof would be reductions in the body's stress hormone, cortisol. Cortisol is produced by the adrenal glands, which sit on top of the kidneys. When a person is stressed for any reason, the brain sends a signal to the pituitary gland, which releases a hormone that flows through the blood to the adrenal gland, causing the release of cortisol. The effects are nearly instant. Cortisol causes blood pressure to rise and the heart to beat faster. These are beneficial effects if you need to jump into action, but if the adrenal gland continues to release cortisol because of chronic stress, its effects will begin to damage the body. Chronically high levels of cortisol cause stomach ulcers, hypertension, and diabetes.

Some studies have found that dogs decrease cortisol levels, while others have not. There is relatively little research in this area, so much of the variability in results probably comes from the variety of conditions in which dog-human interactions have been studied. Not everyone likes dogs, and as Lyra proved at the lab party, dogs can send cortisol levels skyrocketing in people who are afraid of them.

Even though there is not a lot of biological evidence yet to prove that dogs have health benefits for humans, some companies have recognized that their employees are happier and more productive when they are with their dogs. Google, for example, states, "[Our] affection for our canine friends is an integral facet of our corporate culture.

We like cats, but we're a dog company, so as a general rule we feel cats visiting our offices would be fairly stressed out." Amazon has a similar policy, simply requiring that employees register the dog and be responsible for good canine citizenship (barking and peeing are no-no's). Other large companies with dog-friendly policies include Ben & Jerry's Ice Cream, Clif Bar, the Humane Society headquarters, Build-A-Bear Workshop headquarters, and the software maker Autodesk. And, of course, many small businesses around the country.

If having dogs at work makes the humans less stressed, do the dogs feel happier too? The question is embedded in the much deeper riddle of animal emotions and gets to the heart of why we were doing the Dog Project.

For the most part, scientists have ignored the question of whether animals have emotions. This is peculiar because most pet owners are pretty sure that they do. Science, though, deals with things that you can measure, and, by definition, emotions are internal. Science has been able to measure only behaviors that are a *result* of an emotion. With humans, this is not a problem. You can always ask a person how he is feeling and deduce which emotion is associated with a behavior. The linking of subjective states and objective behaviors is an important step because different emotions may result in similar behaviors and expressions. For example, if you see someone crying, you might assume he or she is sad. But those could be tears of joy. The only way to know is to ask.

This inability to exactly determine emotions from behavior is why scientists have generally avoided the question of animal emotions. For example, a dog can't tell you why he chews your slipper. But scientists have not always been so reluctant to venture into this. Charles Darwin devoted an entire book to the topic. In *The Expression of the Emotions in Man and Animals*, Darwin described how emotions like joy and fear have common manifestations in both animals and humans. Although *Expression of the Emotions* was his third book, after his famous books on evolution, it is the one that resonates most strongly today. The timelessness comes from his heavy reliance on dogs to illustrate his points. Richly illustrated with photographs and engravings, the modern reader can immediately identify with Darwin's dogs.

Because humans and animals evolved from a common ancestor, Darwin deduced that we might also share basic emotional functions. If that were the case, animal emotions would help reveal the origins of human emotions. Unlike other scientists of his era, content to simply describe natural phenomena, Darwin wanted to understand why emotions manifested the way they did. Why, for example, does happiness trigger an upturned mouth as opposed to a downturned one?

Darwin formulated three principles of emotions that applied to man and animals. First, he said that emotions come from the brain. This was a pretty remarkable and correct intuition, considering that almost nothing was known about the brain in 1872. Second, emotional expressions build on natural movements. For instance, smiles are upturned because laughter triggers the closing of the eyes, and the contraction of the muscles around the eyes also raises the corners of the mouth. Third, Darwin believed that emotions manifest as the opposite actions of opposing habits. Darwin chose a dog to illustrate this principle, which he called *antithesis*.

When a dog approaches a stranger that appears hostile, the dog "walks upright and very stiffly; his head is slightly raised . . . ; the tail is held erect and quite rigid; the hairs bristle . . . the pricked ears are directed forwards, and the eyes have a fixed stare." These actions are defensive and may represent a prelude to an attack. The principle of antithesis states that the opposite emotion—joy—manifests with opposite motions. "Instead of walking upright, the body sinks downwards or even crouches . . . ; his tail, instead of being held stiff and upright, is lowered and wagged from side to side." The descriptions are as apt today as they were 150 years ago.

Darwin's work on emotions was forgotten for more than a century. Although serious research in this area is beginning to attract scientists again, the vast majority still stays away from the knotty question of animal emotions. A major factor in scientists' reluctance is that the study of animal emotions opens up an uncomfortable ethical question. If animals have emotions like humans, is it right to kill and eat them?

There have been a few exceptions. Within neuroscience, two people stand out. Kent Berridge, a psychobiologist at the University of Michigan, has extensively studied the link between reward systems in the brain and the expression of emotion in rats. And Jaak Panksepp, a neuroscientist at Washington State University and Bowling Green State University in Ohio, has been the strongest advocate for mapping animal emotions onto corresponding brain systems that are common to all mammals.

Reiterating what Darwin said, Panksepp has argued that only when we understand the emotional systems of our fellow creatures will we begin to understand the origins of human feelings. This is a compelling argument. When we look at the brains of animals, it is immediately apparent that there are many structures in common. The commonalities have traditionally been called *primitive*, reflecting scientists' belief that they must have an old evolutionary origin. In the 1960s, the neuroscientist Paul MacLean used the evolutionary analogy to divide the brain into three parts: the reptilian brain (the basal ganglia), the paleomammalian brain (the limbic system), and the neomammalian brain (the neocortex). Although these divisions are overly simplistic, it is clear that only the neocortex is substantially different in humans and other mammals. The other two divisions— the basal ganglia and the limbic system—are largely the same from rats to humans. It is in these systems that Berridge and Panksepp believe that emotions originate.

The first difficulty in studying animal emotions lies in describing what an emotion is. Humans have a rich language for emotion, but even if you take something as basic and universal as love, you quickly realize the vast nuances that that word contains. There are so many different types of love that the word itself is inadequate. Assuming, for the moment, that our dogs love us, what kind of love would that be?

To proceed scientifically, we must set aside such subtleties. It helps to break emotion down to fundamental components: valence and arousal. Valence is simply goodness or badness, while arousal describes the level of excitement, ranging from calm to maximum excitement. Many human emotions can be plotted on a graph as a function of the combination of valence and arousal. Because the graph forms a circle, it is called the *circumplex* model of emotion. Positive emotions are plotted to the right, while negative ones are on the left. In the vertical direction, high-arousal emotions are at the top, while low-arousal ones are at the bottom.

Many psychologists have argued that the two-factor model is too simplistic. However, it provides an excellent starting point to understanding which parts of the brain give rise to the different emotions. As it turns out, the reptilian part of the brain, which we now call the basal ganglia, is closely associated with positive valence, while the limbic system is associated with arousal. By examining the relationship of activity in these different brain systems to the emotions experienced by human subjects, we can build an emotional brain map.

Because dogs have basal ganglia and limbic systems that look almost the same as ours, such a map could be applied to dog brains to help determine what a dog is feeling.

In the upper left portion of the circumplex are the emotions with high arousal and negative valence. The usual behavioral manifestation of these emotions is avoidance or retreat from whatever caused them. However, the close proximity of emotions like fear, anger, and frustration in the circumplex make that quadrant difficult to map in the brain. Despite having similar levels of valence and arousal, those emotions feel quite different from one another. Although much is known about the fear system of the brain, almost nothing is known about rage or frustration.

The upper left quadrant would remain uncharted territory in the Dog Project because of the ethical problems with inducing those types of emotions in our dogs.

In contrast, the upper right quadrant of the circumplex model is well understood and seemed like an excellent place to begin mapping the dog brain. These are the emotions that are maximally enjoyable: very good and very exciting. These positive emotions are also associated with a specific behavior seen in all animals. If something is good and exciting, every animal—dog, rat, human—will approach it. Panksepp calls this the *seeking* system. In the brain, we know that approach behavior, as well as the corresponding positive emotions, is associated with activity in a tiny part of the basal ganglia called the nucleus accumbens. In humans, when we observe activation in this region we can deduce that the person is experiencing a positive emotion and very likely wants whatever is making them excited.

Although I couldn't know for sure, when Callie saw me with the bag of hot dogs, her nucleus accumbens was probably lighting up like a Christmas tree. Wagging her tail and running toward me was a classic approach behavior. This led me to assume that she was experiencing joy and excitement. But only through fMRI would we know what she truly felt.

13

The Lost Wedding Ring

UNLESS WE CAME UP WITH a solution for the head movement problem, the Dog Project would grind to a halt. We had two choices: either train the dogs to hold their heads still or come up with a better chin rest.

Mark was confident we could train the dogs. Considering that we needed the dogs to move less than two millimeters in every direction, this struck me as a significant obstacle. As I worked with Callie, I couldn't even discern movements that small. She could easily move a few millimeters while I was looking away, and I wouldn't even know. The alternative, a more restrictive chin rest, didn't appeal either. I didn't want to encase the dogs' heads in plastic like what Andrew and I saw at Yerkes.

We were at an impasse.

Movement *during* a scan causes ghosting, but movement *between* scans causes a different problem. At the beginning of a series of scans, we have to set the boundaries of the scan, called the field of view (FOV). As Callie demonstrated during the dress rehearsal, she didn't always place her head in the same position in the head coil. Because of the inconsistency of her head placement, sometimes she was in the

FOV, but most of the time she wasn't. All but one of the images were empty.

We had a few tricks we could do with the functional scans that would help with the movement problems. An easy fix for ghosting is to shorten the time it takes to acquire an image of the brain. By shortening the scan time, it makes it less likely the subject will move during that period. But the only way to shorten the scan is to take fewer slices. Each slice through the brain takes roughly sixty milliseconds. For humans, it usually takes thirty slices to cover the whole brain, for a total of two seconds. Fortunately, dogs have smaller brains than humans, so we don't need as many slices. But if we get fewer slices, the FOV shrinks, and the between-scan movement becomes a bigger problem. With a small FOV, we might miss the brain entirely if the dog didn't put her head in the correct location.

We needed a solution to both the within-scan and between-scan motion.

The problem was the chin rest. The foam bar provided feedback to the dogs about where to place their heads in the up-down direction, but left them free in the left-right and forward-backward directions. We needed something that would guide the dogs to place their heads in exactly the same position every time they went into the coil and would keep them there for the duration of the scan.

Atlanta had caught a week of unseasonably warm weather that January. It was a good opportunity to get outside, and I figured a change of scenery might generate some new ideas about the movement problem. So Kat and I piled the girls and the dogs into the minivan and headed down to the river for some hiking.

The Chattahoochee River originates in the northeast corner of Georgia, in the foothills of the Blue Ridge Mountains. From there, it flows southwest toward Atlanta, picking up volume along the way. Eventually, the "Hooch" flows all the way to the Gulf of Mexico. Much of the area around the Chattahoochee is national forest, and we were fortunate to live only a mile from the river. It was a great place to hike or mountain bike or just relax on the banks and watch wildlife.

Our favorite spot was Sope Creek — a rushing stream that tumbles down to the river over a series of granite boulders. In the decade before the Civil War, a ferry shuttled people and lumber across the Chattahoochee where the creek met the river. Next to the creek, the ruins of an antebellum paper mill had been overtaken with kudzu, the aggressive southern version of English ivy. We spread out our towels beneath the ruins while Helen and Maddy jumped from rock to rock across the creek. It had rained a few days earlier, so the flow was higher than usual.

Lyra, despite her golden retriever genes, would wade into the water only until it touched her belly. Since she had no interest in swimming, I let her off-leash. She just milled about, sniffing rocks and plants and nosing around for bits of food other people had left behind.

I looked at Callie. Her prey instinct was on red alert. She sat stiffly at the end of the leash, head like a periscope, twisting in lightning-fast jerks toward every sound and motion in the woods. She looked up at me and whimpered. I didn't need an MRI to know what she wanted. She wanted to be off-leash like Lyra.

Figuring she would hang around the picnic spread that Kat had been setting up, I reached down to unclip her. The world seemed to spin down. As I pulled back on the leash clip, Kat screamed out in slow motion "Nooooo!"

With the opening of the clip, everything kicked into high speed. With that telltale *click*, Callie knew she was free.

She never looked back. Like a cheetah, Callie arched her back and released all of her pent-up energy into a massive sprint. Until that moment, I'd had no idea she could swim. But what it really looked like was a twenty-pound stone skipping across the creek. In three bounds, before anyone could react, she had hopped across the water and disappeared into the woods on the other side.

"Dad!" Helen screamed. "Callie is running away!"

I jumped into the creek. The current was strong, so I faced upstream and crab-walked as fast as I could, grabbing on to boulders for support. It must have taken at least a minute to wade across, and I could mentally tick off the distance that Callie was covering with each second.

The other side of the creek was thick with kudzu and poison ivy, but a thin path headed downstream toward the Chattahoochee. If Callie made it that far, her next stop would be the Gulf of Mexico. I couldn't bear the thought of losing another dog so soon. Never mind that the only dog in the world that had voluntarily sat for an MRI, even if only a dress rehearsal, had just escaped because of my own stupidity.

Kat pulled herself up the bank right behind me and we both set off down the footpath, yelling, "Callie, wanna treat?"

I ran ahead of Kat. Not fifty yards away, I couldn't hear even her behind me; the sound of rushing water from the creek bounced off the trees and seemed to come from everywhere. There was no way Callie could hear us. A sense of panic started rising in my chest. I couldn't go back to Helen and Maddy without Callie.

The path came to an end at the edge of a side stream feeding into Sope Creek. With nowhere else to go, I slid down an embankment and waded back into the creek. Hopping from rock to rock, I made my way slowly downstream toward the Hooch. Kat was at least a hundred yards back. I continued to call out for Callie, but even if she was nearby, I doubted that she heard me.

A couple was sunbathing on a rock in the middle of the creek.

"Did you see a little black dog?" I asked.

They shook their heads.

The search operation had now been going on for ten minutes. Callie was nowhere to be seen. I tried to listen for the jangling of her dog tags, but all I could hear was the water and the laughter of a group of teenagers a little farther down the creek.

Keep going or turn back? Callie was in unfamiliar territory. There was no way to know what she would do. Given her fascination with squirrels and chipmunks, she could have gone anywhere in the forest.

The teenagers' laughter got louder. Foolishly, I resented it. I had just lost another dog, and it was all my fault. What was there to laugh about?

Then I saw why the teens were laughing. They were pointing at something moving at high speed through the creek.

It was Callie. She was half jumping and half swimming, in hot pursuit of a gaggle of young geese.

"Callie!" I screamed. "Come here, girl!"

She didn't hear me.

The geese were not particularly coordinated and couldn't fly very well. The whole group was moving erratically from one side of the creek to the other. And then the geese turned to backtrack upstream.

They flew by, barely above the water. A second later, Callie screamed by, just out of reach. With her attention focused on the geese, she didn't even notice me.

"Callie!"

The scrum was headed for Kat.

Kat made a leap to grab Callie, but her timing was off, and Kat splayed out on a rock, empty-handed.

I started making my way upstream again. It would have been faster if I could get to the bank and run up the path, but I would risk missing Callie in the water. I could make out her splashing about a hundred yards away, still in pursuit of the geese. Kat was wading through the rapids but not making much progress either.

"She's coming back!" Kat yelled.

For reasons only the geese knew, they had decided to make another 180-degree change in course. Four fluffy goslings were flapping straight at me. This was going to be my last chance.

Not wanting to frighten the geese into another course change, I remained motionless in the center of the water. They whizzed by in a chorus of discontent. The bounding water dog was a few seconds behind them, oblivious to anything but the chase.

As she closed to within ten yards, I called out in my happy voice, "Callie! Good Girl!"

It was enough to momentarily divert her attention from the birds. She glanced over at me. I lunged out and snagged her collar with a finger.

Drawing her in for a hug, I gloried in the smell of wet dog.

"Callie, you crazy dog. We almost lost you!"

She let out a heavy sigh as she watched her prey vanish downriver.

I carried her back to the bank and towed her to where the girls

Callie after shootin' the Hooch.
(*Gregory Berns*)

were waiting. Helen immediately buried her face in Callie's slicked-back fur.

"Dad!" she exclaimed. "Don't ever do that again!"

Kat pulled up, holding her hand.

"What happened?" I asked. She held out her left hand. Her ring finger was bent at an awkward angle. She had dislocated her finger trying to catch Callie. It was beginning to swell.

"I have to take off my wedding ring," she said, "or it might have to be cut off." I didn't know it at the time, but it would be a year before that finger returned to normal size and she could wear her ring again.

I was relieved that we had Callie back. The episode made me realize that despite all of our high-tech training for the MRI, I still didn't have a clue as to what she was thinking.

By the time we got home, Callie had already shrugged off her trip downriver, and the kids had taken to exercising their creativity on the stack of boogie boards in the basement. The nearest ocean was three hundred miles away, so something else had to substitute for water. It didn't take them long to figure out that sliding down the carpeted steps was a hoot.

Helen and Maddy laid out pillows at the bottom of the steps. In rapid fire, they placed a board at the top of the stairs and went shooting down into the pillow pit. Of course, this was highly exciting to Callie and Lyra. Callie zoomed up and down the stairs trying to catch the kids as they went tumbling past. As Callie chased the girls, she made a *huuuf-huuuf-huuuf* sound that sounded a lot like a person trying not to laugh. Lyra just stayed at the bottom of the stairs, barking incessantly. A strand of drool stretched to the floor.

When the girls got tired of stair surfing, they used the boogie boards as shields while play jousting in the basement. When they got tired of that, they just started whacking each other.

"Daa-ad!" someone yelled. "She broke one of your boogie boards!"

I ventured into the melee. Sure enough, one of the boards had snapped into two pieces. Callie was in a sphinx position on the floor, swishing her tail, and looked up at me with an I-didn't-do-it expression on her face. Lyra, who was chewing on one of the chunks, had taken a bite out of one of the pieces and was about to swallow it when I reached into her mouth and swept it away. I picked up the board and stared at the crescent she had made. It was about the size of Callie's chin.

A lightbulb went off in my head.

I retrieved a utility knife from the garage and cut the broken boogie board into strips. Carefully, I began carving out a semicircle on the edge of each piece. After each pass, I would check the fit against Callie's chin. The first piece had a small cutout for the end of her muzzle. The next piece had a bigger cutout, and the next bigger yet. Stacked against each other, the cutouts were beginning to form a three-dimensional cradle. The fourth piece was the biggest. A deep cutout allowed the foam to extend up to Callie's ears and fit behind the back of her jaw. This provided a secure support both up and down and, crucially, forward and backward.

With the sandwich of four pieces taped together, I checked the fit on Callie. She had retired to the sofa. Gently, I lifted up her head and placed the foam sandwich beneath her chin. She relaxed and looked at me with indifference.

I was ecstatic. I snapped some photos to send to Andrew and Mark. The new chin rest would solve our motion problems. It was firm, so it would support the weight of Callie's head and prevent movement up and down. But the cutout also provided her with positive feedback on where to place her head. Her chin would fit in only one way, and the cutout guaranteed that as long as Callie nestled her head down into it, the location would be the same left to right and front to back.

Callie testing the fit of the boogie board chin rest.
(Gregory Berns)

The next day, I trimmed up the foam block and glued it to a piece of plywood that would span the diameter of the head coil. The plywood was cut to just the right length so that the whole chin rest placed Callie's head in the center of the coil while allowing space beneath it for her paws to stick forward.

With the contraption on the floor, I called to Callie, "Coil!"

She scooted in and immediately plopped her head in the rest.

"Excellent!" I exclaimed, and gave her a piece of hot dog. She happily swallowed it and placed her head back down. She waited for more.

Next, we tried it with the earmuffs. Callie still didn't like them, but with enough hot dogs, I coaxed her into the coil while wearing the muffs. I tapped the new chin rest, and she dutifully placed her head in the cradle. The muffs slipped back a little, but I was able to slide them forward again over her head. Callie's eyes dilated as she anticipated treats.

"You are such a good dog!"

Callie appeared so comfortable in the new chin rest that I began lengthening the time she had to stay motionless before giving her a treat. In a dozen repetitions, she was holding absolutely still for up to ten seconds at a time. That was more than enough. We would be able to get at least five full scans of her brain in that period.

The new chin rest design looked very promising. I had already booked the MRI for Dog Day, which was now less than two weeks away. The scanner was reserved for four hours to allow us plenty of time to let the dogs get comfortable in the scanner, but at $500 an

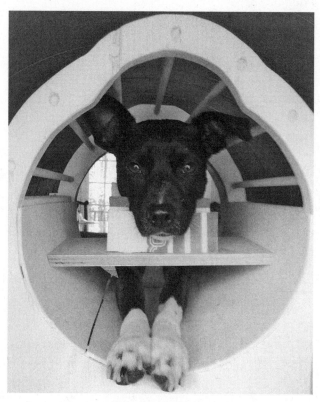

The full monty: Callie in the final version of the chin rest.
(Gregory Berns)

hour, this was going to be an expensive experiment. Of course, we wouldn't know if it was successful until we actually tried it at the scanner. The new chin rest was the best shot we had, so I made one for McKenzie too.

Both dogs made rapid progress with the new chin rest. McKenzie, of course, was already a champ at holding still. The new rest also made it easy for the dogs to consistently place their heads in the same location. There was only one last detail to work out before scan day.

What, exactly, was the scientific question we hoped to answer by scanning dog brains?

14

Big Questions

WITH THE NEW CHIN RESTS, Callie and McKenzie were cruising along in their training. I was routinely blasting the scanner noise at 90 decibels, and as long as Callie had the earmuffs on, she didn't seem to mind. Scan day was two weeks away. Andrew had circled the date on the lab calendar and written "Dog Day" in bold red letters.

None of us expected to pull off a complete scientific experiment on the first try, so we held on to modest expectations. The first and most important goal of the scan session would be the acquisition of a sequence of fMRI images that weren't contaminated by motion artifacts. I would consider the session a success if we obtained ten images in a row without the dog moving. That would mean holding still for a twenty-second interval in the scanner. Considering how well the dogs were doing in their training, that seemed entirely possible.

On the off chance that the dogs surpassed our expectations and miraculously held still for several minutes, we would then have the opportunity to collect enough data to go beyond simply proving the viability of the Dog Project. We might actually get to answer a scientific question about canine brain function, which, of course, was the whole point. I didn't expect to be able to do this on the first go, but it's always best to be prepared.

I called a meeting of the core Dog Project team — Andrew, Mark, and me — but since everyone in the lab was rooting for this, it turned into an impromptu lab meeting.

"We've got two weeks until Dog Day," I began, "and we have to nail down the experimental task that the dogs will do."

Because the scanner noise was so loud, the dogs would not be able to hear vocal commands. That left hand signals as the primary means of communication while in the scanner. Up until now, we had purposely avoided using hand signals because we didn't know what kinds of signals we would use and what the dogs should do with them. It was time to figure that out.

"Not much is known about the functional organization of the dog brain," Andrew said. "We don't even know what parts of the dog brain are responsible for basic functions like vision and hearing." What we did know came from some unsavory experiments over a century old. In 1870, a pair of German scientists used the then new technology of electrical energy to directly stimulate the brains of animals. By sticking their electrodes in different parts of the brain, they discovered that the electricity could cause an animal to move its limbs. Sadly, they used dogs for this experiment, even puppies — a trend that continued into the 1970s. The end result of this research was the knowledge of which parts of the dog brain controlled movement. But for the Dog Project, we couldn't study the parts of the brain that controlled movement because the dogs weren't supposed to move!

"Let's go with our strength," I said. "Our lab has spent the last decade studying the human reward system. We know a lot about how it works. There isn't any reason to expect the dog's reward system to be any different from the human one."

"Reward-prediction error experiment?" Andrew asked.

The brains of all animals appear to act as prediction engines. Prediction, after all, is key to survival. If your brain weren't predicting what would happen next, you wouldn't be able to walk across the street without being hit by a car. Most of the brain's predictions have to do with things in our environment, like cars, and things that other people are doing. The caudate nucleus (located within the basal ganglia) and the parts of the brain that feed into it are concerned with predicting rewards.

In the early 1990s, Wolfram Schultz, a Swiss neuroscientist, measured the activity of neurons in monkeys' brains while they were trained on a simple classical conditioning task. When a light turned on in a monkey's cage, it received a squirt of fruit juice in its mouth. Just like Pavlov's dogs, the monkeys quickly began to anticipate the juice when the light came on. Schultz discovered that the neurons in specific parts of the brain followed the same pattern. Initially the neurons fired in response to the juice, but once the monkeys had learned the association with the light, the neurons fired to the light, not the juice. The neurons that showed this pattern were located in the heart of the reward system, the caudate.

Since Schultz's discovery, neuroscientists have learned that these neurons don't signal things that are pleasurable. Instead, they fire when something unexpected occurs that indicates something good is *about to* happen. If something is unexpected, then that means the brain made an "error" in predicting it. For this reason, scientists call these events *reward-prediction errors.*

We know where reward-prediction errors occur in the brains of monkeys and humans. Dogs, we figured, should be no different. And because the caudate is a well-defined structure, it made sense that we would be able to identify it in the dogs' brains — assuming, of course, they held still for the MRI.

"We could train the dogs with a hand signal that indicated they would receive a hot dog," I said.

"If the dogs learned the association between the hand signal and the treat," Andrew agreed, "we should see caudate activity to the hand signal."

"Just like Schultz's monkeys," I concluded.

Lisa spoke up and pointed out a flaw in this reasoning.

"How would you know that the dogs had learned the hand signal?" she asked. "After all, they're not doing anything."

She had a point. All of behaviorist learning theory depended on the manifestation of either a response, like drooling, or a behavior to indicate that the animal has actually learned something. We would have only the brain.

"We'll have to rely on the dogs' caudate," I said. "A response there would be proof that they learned the signal. We could also look for other signs, like the pupils dilating in anticipation."

There was another problem. An fMRI scan measures brain activity indirectly. What it actually measures are changes in the oxygen content of tiny blood vessels in the brain. When neurons fire, the surrounding blood vessels expand a little and let in more fresh blood for the neurons to replenish their energy storage. The scan picks up these changes in blood flow, and from that we deduce which neurons were active. But there is a catch. The brain is always on. It is a myth that we use only some small percentage of our brains. The truth is that we use all of it — just not all at once. Because the brain is always on, and blood is always flowing, fMRI can measure only *changes* in activity. When designing fMRI experiments, you always need a comparison, or baseline, condition.

Callie would be in the scanner, holding her head still and watching me. So many things could be going on in her brain there might not be a way to interpret the fMRI measurements. Even if we trained the dogs on a hand signal, we would still need something to compare their brain responses to. Ideally, the comparison condition would be almost the same as the thing of interest. You want to keep everything the same in both conditions except for the one thing that is being varied in the experiment.

To measure the response to a hand signal, we needed another hand signal as a comparison condition. This way, everything would be the same — holding still, watching the handler and even the handler's movements. We would vary the *meaning* of the signals.

"How about another hand signal," I suggested, "which means something else?"

"Like what?" Andrew asked.

"A different type of food," I said. "Something the dogs don't like as much as hot dogs."

"Like what?" Lisa asked. "Sheriff likes everything."

It was a fine line. We wouldn't want the dogs to eat something nasty. We needed something that they would eat but not like as much as hot dogs. Dogs are mostly carnivores. It seemed logical that they wouldn't value a vegetable as much as a piece of meat.

"How about peas?"

Everyone nodded as they envisioned how this would work. I held up my left hand in a "stop" gesture.

"Suppose this means 'hot dog.'" I thought briefly about holding up my right hand for "pea," but as we didn't know the extent to which dogs distinguished left and right, this seemed like a bad idea. Instead, I held both hands flat in front of my chest, pointing toward each other. "And suppose this means 'pea'?"

Mark nodded.

"Those signals should be easily distinguishable to a dog."

The rest of the team agreed.

It was decided. The first canine fMRI experiment would be "Peas versus Hot Dogs."

Over the next week, Andrew and I formalized the design of the experiment, which is in some ways like writing a screenplay. Every detail has to be planned in advance. The lab walls became our story-board. We needed to decide how many times we would give peas and hot dogs and the order of their presentation. Dogs are very good at learning sequences of things, so we wouldn't want to simply alternate between peas and hot dogs. If we did, the dogs would know that as soon as they got a pea, the hot dog would be coming next, and there would be no need to pay attention to the hand signals. To prevent this, the order would have to be random.

The most important detail, though, would be the timing of the experiment. Each repetition would have four elements. First, the dog would place her head in the chin rest. Because of the associated movement, this would cause artifacts on the scan being acquired at that moment. We would need to wait at least two seconds for the next scan to begin. Once the dog was settled in the chin rest and enough time had passed for the artifacts to decay, we would proceed to the second element, the hand signal.

Melissa and I would be giving the signals to our dogs, and all of our attention would be focused on Callie and McKenzie. It would be too much for us to randomly decide on the fly which hand signal to give, so Andrew would be standing next to us with a pregenerated list of the order of signals. He would hold up one finger for hot dogs and two for peas. The handler, facing the dog in the scanner, would then give the corresponding hand signal. Timing was critical.

We knew that the fMRI response would not be instantaneous. The blood vessels surrounding the neurons take a few seconds to dilate, peaking after six seconds and returning to baseline twenty seconds after that. This profile is called the *hemodynamic response function*, or HRF, and it is a bugaboo of fMRI experiments. The lag in response meant that the dogs would have to hold perfectly still for the time it took the HRF to peak and decay. This delay period was the third element. Ideally, the dog would hold still for twenty seconds. But I would settle for ten, which would be enough time to capture at least the peak of the HRF.

Mark and Andrew and I debated whether we should just flash the hand signal for a second and then wait ten seconds before giving the reward. The alternative would be to hold up the hand signal for the entire waiting period. In the end, we opted for the latter. If we flashed the signal briefly, we wouldn't know whether the dogs were paying attention to the appearance of the hand signal or its removal. Both could be salient. To make sure that they paid attention to the signal's appearance, it seemed prudent to keep our hands up until we gave them the treat, which was the fourth element and the end of each sequence.

This would be easy for the hot dog repetitions. Left hand up for the signal, and then deliver the hot dog with the right hand. The pea repetitions, though, required both hands as part of the signal. This meant that the pea would have to be palmed in the right hand.

To make sure we were smooth on scan day, Melissa and I practiced all the elements over and over again. The dogs might pick up subtle changes in our body language, but there wasn't much we could do about it other than practice.

At home, Callie and I continued working with the mockup. She was so used to it, even with the scanner noise at 95 decibels, that she was starting to get bored with the routine. She liked the hot dogs, but once inside the head coil, it was all business. No tail wagging anymore. Just a look that said, *I'm here waiting for my treats.* The day before scan day, we practiced for barely ten minutes. No sense in wearing her out.

Callie was ready. If she knew what was about to happen, she gave no sign.

I, on the other hand, could barely contain myself. We had made a lot of progress since the dress rehearsal. The redesign of the chin rest, the modification of the scan parameters, and the choreography of the experiment were all positive developments. But still, there was a lot riding on this. Tomorrow, we would know for sure whether the Dog Project would work.

Helen had been eagerly waiting for Dog Day too. She had made an earnest attempt to improve her study habits, making flash cards and going through them diligently. The night before scan day, I kissed her good night.

"Dad," she asked, "am I going with you to work tomorrow?"

"Yes," I said. "I'm very proud of how hard you've been working. I know science is a hard class. Tomorrow you'll see how much fun science can be."

She smiled and gave me a hug.

There was no sleep for me that night. I lay in bed wide awake. Callie was curled up between Kat and me. I rested a hand on Callie's smooth fur and immediately felt the calming effect of her chest rising up and down slowly. Newton had had the same calming effect, augmented by his soft snoring.

With nothing to do but let my imagination spin away, Edward Jenner popped into my head. In 1796, Jenner invented the smallpox vaccine. He had made the astute observation that women who milked cows were immune to smallpox. Jenner suspected that the blisters the milkmaids acquired from a similar disease, cowpox, contained a substance that could provide immunity to other people. Jenner tested his theory on James Phipps, the eight-year-old son of his gardener. After inoculating James with pus from a milkmaid, Jenner exposed him to the real smallpox virus. James didn't get sick. Thanks to Jenner, the world is now rid of smallpox.

What Jenner did, though, could never be done today. He took an outrageous risk. If he had been wrong, the boy would have contracted smallpox and would probably have died. Perhaps the gardener had no choice in the matter, but I still admire Jenner for having the courage to test his theory on a member of his own household. If today's biomedical researchers were required to test their theories first on people they know, there would be a lot less crap making it into the scientific archives.

Callie was a part of our family. And I was about to pull a Jenner on her. I had no qualms about going into the MRI myself. I routinely volunteered for the lab's experiments. But Callie wasn't human. There was much we didn't know about dogs. Mark had told me stories about dogs that had found their way home after being lost hundreds of miles away. How did they do that? Maybe they were like pigeons and had some primitive magnetic sense in their brains. Would the MRI blind Callie's sense of direction?

We were about to venture into unknown territory. People would question what we were doing. Some might view this as animal torture, even though we had elevated the dogs' rights to those of humans.

There was no other way. It had to be family.

15

Dog Day Afternoon

THE NEXT MORNING, HELEN HELPED me load up the car. I couldn't tell if she was more excited about seeing the experiment or playing hooky from school. It didn't matter. It was great to have her along.

I'd prepared a checklist to make sure we didn't forget anything: earmuffs, chin rest, hot dogs, peas, nylon collars and leashes with the metal removed to be MRI-safe, and plastic steps. We tried to be discreet about it, but as soon as Callie saw the chin rest, she started pogo-sticking in excitement. She followed us to the garage door and wiggled between my legs to run to the car.

The three of us arrived at the lab just before noon, and it was already packed. Since this might never happen again, we also had a photographer come to document everything. To make sure we were in compliance with the IACUC protocol and to look out for the welfare of the dogs, we had requested a veterinary technician. I had no idea who would be sent and what he or she would think about the Dog Project, but any concerns I'd had instantly disappeared when Rebeccah Hunter introduced herself. Rebeccah was young and en-

Rebeccah being greeted by McKenzie.
(Bryan Meltz)

thusiastic and, most important, a dog person. Callie ran over to her and jumped up to lick her face. As any dog person knows, this is a crucial test in evaluating someone's character. Do they back away in disgust or do they lean into the doggie kiss?

Rebeccah not only leaned in, she knelt down to Callie's level and cooed, "Oh, what a good girl!" Callie planted one right on her lips.

I didn't know it yet, but Rebeccah's rapport with the dogs would be crucial in just a few hours.

Mark and Melissa soon arrived with McKenzie, and we let the dogs run around the lab to burn off nervous energy. They zoomed from person to person, making sure everyone got a good sniffing. Callie's tail never stopped wagging. Even when lying on the floor, her tail would begin thwack-thwacking whenever someone approached

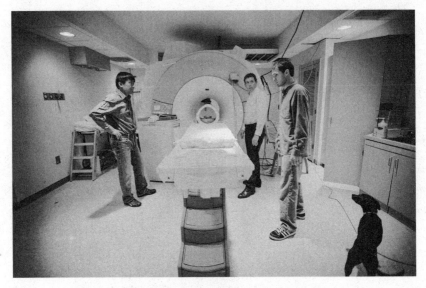

Sinyeob, I, Andrew, and Callie ponder the MRI.
(Bryan Meltz)

her. After ten minutes of playtime, both dogs had settled down. For what was about to happen, we wanted the dogs as calm as possible. Being a little tired would help them hold still in the MRI.

We swapped the dogs' collars for nylon ones. I had replaced the standard metal D-rings and clips on these leashes with plastic equivalents. Everyone did a double-check for metal in their pockets, like keys and cell phones, and for credit cards that would be erased if they got too close to the magnet. Normally, we would do the metal check in the control room at the scanner, but in order to avoid parading the dogs through the halls of the hospital, we would be entering the MRI room directly from outside through a side door. We needed to make sure everyone who was going to be in the room was MRI-safe.

At the scanner, we were greeted by Robert Smith, the tech who would run the MRI and who had been there for Callie's dress rehearsal, and Sinyeob Ahn, a magnetic resonance physicist from China who would tweak scanner settings for us on the fly. They both smiled when they saw Callie and McKenzie, but I could tell neither of them thought this would actually work.

I heaved shut the vaultlike doors to the magnet room, creating a seal impenetrable by any form of electricity or radio waves. With the room secured and the doors locked, we let the dogs off-leash. Callie, of course, had been here before and knew that crumbs of food could be found on the floor of the control room. She gave only a passing glance at the magnet as she went off in search of something to eat. McKenzie gave the MRI her best border collie stare. Despite the susurrations of the cryogen pump, McKenzie soon realized that the magnet was not alive and could not be herded.

While Melissa worked with McKenzie to get comfortable around the magnet, I settled in with Robert and Sinyeob to plan the afternoon's scans. Based on what we had learned from Callie's dress rehearsal, we would implement several changes in the scan protocol.

From the control room I could see Melissa coaxing McKenzie into the scanner. Tentative at first, McKenzie carefully walked up the steps and plopped down on the patient table. She seemed unsure of what to do. Melissa went around to the other end of the magnet and, using some hot dogs, lured McKenzie in. As soon as Callie caught sight of that, she ran up the steps and climbed over McKenzie to stake out her position inside the magnet.

Sinyeob started laughing. "Look, two for one!"

McKenzie was not amused.

Turning to Helen, I said, "Could you get Callie and keep her out of the magnet until we're ready?" Helen beamed, happy to have a job to do.

As had been done at the dress rehearsal, Sinyeob programmed three types of scans. The first would be the localizer image to see where the dog's brain was. This would allow us to set the field of view for the subsequent scans. Next would come the functional scans.

"How long do you want to scan?" asked Robert.

For human studies, we normally scan in blocks of about ten minutes. That is about as long as a person can stay engaged in a task while in the scanner. We usually give the person a break of a few minutes and then repeat. I didn't expect the dogs to stay in the scanner for that long, though.

"Let's go with five minutes," I said. With a complete scan through the brain taking two seconds, that would give us 150 images. Our goal was to get a consecutive sequence of ten images without movement somewhere in that block.

If we had time, and the dogs were up for it, the final scan would be a structural. It would be quick, but they would need to hold still for thirty seconds.

I turned to Andrew and said, "Are you ready?"

"Let's do this!"

I looked at Callie and said, "Hey, Callie! Wanna do some training?" She jumped up to lick my face.

Melissa put McKenzie in a pup tent that she had brought. This let her relax while we worked with Callie.

Rebeccah followed Callie and me into the magnet room and took up a position at the end of the patient table where she could closely monitor Callie for any signs of distress. I slipped the earmuffs over Callie's head and motioned for her to go into the magnet. As she shimmied in, I ran around to the back end so that I was facing her. Callie scooted into the head coil and placed her head on the chin rest.

"Good girl!" I said, and gave her a hot dog. She wagged her tail and looked at me expectantly. Meanwhile, Andrew had taken up a position just to my left. I glanced at him.

"Do it."

Andrew raised his hand and pointed to Robert, who was observing through a glass window in the control room. The scanner came to life. *Click. Click. Click.* Then the buzzing of a million bees. The software, not recognizing a canine occupant, was beginning the shimming procedure to compensate for the distortion of the magnetic field that Callie's canine form was causing.

Callie's eyes narrowed to slits. I held up a piece of hot dog but it was already too late. She had started to back out. Further confused by the lack of anything inside it, the scanner continued buzzing for a few seconds until it gave up and aborted the sequence.

Rebeccah consoled Callie by stroking her chest. The earmuffs had slipped back and dangled uselessly around Callie's neck.

I gave her some more hot dog pieces and repositioned the earmuffs to try again.

Callie went inside and once again assumed the sphinx position in the head coil. And once again, as soon as the scanner started shimming, she backed out.

After two more tries, I was beginning to get frustrated. I sat on the doggie steps and petted Callie on the head. She just looked at me as if to say, *I'm trying*.

Even with the extra training, the noise was worse than we had anticipated. Plus, the earmuffs kept sliding off. Maybe if we could get the earmuffs to make a better seal and stay on, the noise would be tolerable.

Rebeccah had the same thought. She rummaged through her vet tech gear and pulled out some gauze pads and a roll of nonadhesive tape — a clingy material called vet wrap. While I fed Callie a constant stream of hot dog, Rebeccah carefully placed a gauze pad between each of Callie's ears and the earmuff. To keep it in place, Rebeccah wrapped her whole head with the vet wrap. Amazingly, Callie didn't seem to mind this procedure. She ended up looking like she'd sustained a serious head injury, but the earmuffs were secure.

"Let's try it again," I said. Callie didn't hear me, which was a good sign. I just motioned to the magnet, and she trotted in.

The magnet went through its click-whirrings, and I braced myself for the swarm of bees. I gave Callie some hot dog, and this time she

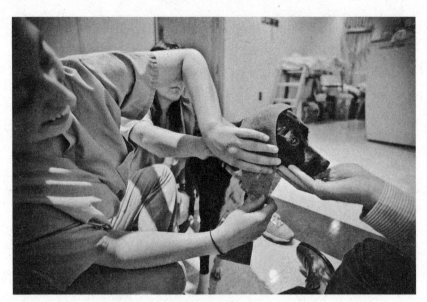

Rebeccah giving Callie the wrap treatment.
(Bryan Meltz)

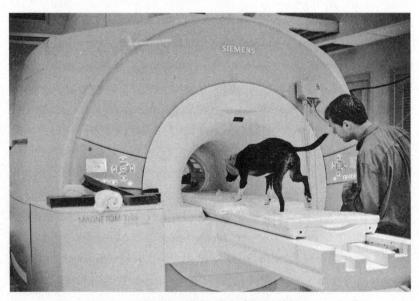

Callie wrapped and ready to rock!
(Bryan Meltz)

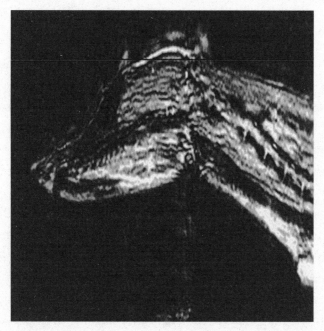

First localizer image of Callie.
(*Gregory Berns*)

stayed put. It seemed to go on forever. After thirty seconds, the buzz-ing was replaced by three brief klaxon sounds of the localizer. That meant the scanner had successfully shimmed the magnetic field and acquired an image. I gave Callie some more hot dog and ran to the control room to see what we had gotten.

Robert and Sinyeob were smiling and giving the thumbs-up sign through the window.

Sure enough, there on the scanner console was a blurry but clearly recognizable image of Callie in profile. Her brain and spinal cord were unmistakable. Everyone gazed in amazement at what was surely the first MRI image of a dog in a natural position. Every other image like this had been obtained in anesthetized dogs who had been in-tubated and who had their necks hyperextended unnaturally. It was eerie seeing Callie's brain transition into her spinal cord and how it ran down her neck just behind her trachea.

Next up were the functional scans. Robert opened a box on the screen to set the field of view. We oriented the FOV so that it was as if Callie's head were a loaf of bread; the MRI would digitally slice her brain face-on, in what is called the *coronal* plane. Each slice would be 3.5 millimeters thick and, with twenty-five slices, that meant the depth of the box would be 87.5 millimeters — just under 3.5 inches. It didn't leave much margin for error. If Callie placed her head in a different location, her brain would be out of the FOV, even if she held it perfectly still. I hoped the chin rest would do its job.

I looked at the clock. It was three o'clock. We had already burned up two hours just to get to this point, and we hadn't even done anything with McKenzie yet. I made a mental note to switch to McKenzie at three thirty.

Callie's energy level had noticeably decreased. She trotted up the stairs into the magnet, but now she wasn't wagging her tail as much. This was actually a good thing. When she wagged her tail in the magnet, the fishtailing motion caused her head to move in rhythm with her tail.

Andrew took up his position at the rear of the magnet. He held a small box with four buttons. The button box was connected to a computer in the control room. Each time Andrew pressed a button, the computer would log the exact time after the beginning of the scan. One button represented the beginning of each trial, when Callie placed her head in the chin rest. Andrew would push a second button when I put up the hand signal for hot dogs, and he would push a third button when I gave the signal for peas. A fourth button would be pushed when I actually delivered the reward. By logging the time of each of these events, we would be able to match them up to the corresponding scan number.

Callie had settled in and looked at me expectantly. I gave her a hot dog and yelled, "Good girl!" I nodded to Andrew to indicate she was ready.

The scanner emitted a few clicks and then launched right into the jackhammer sounds of the functional scans. This was the sequence Callie had been practicing with for the last month, but it was much louder than the localizer. She lifted her head up from the chin rest and started to retreat, but then she paused. Half in and half out, she stared at me. I held up a piece of hot dog. Callie thought about this for a second, and then scooted forward to lick the treat out of my hand. Satisfied that nothing bad was going to happen, she settled into the chin rest.

Out of the corner of my eye, I could see Andrew hit the first button. He was holding up one finger — the sign for hot dogs.

In my head, I counted two Mississippis and put up the hot dog sign. Callie's eyes dilated. I tried to count another six Mississippis before giving her the treat. It was just like practicing at home.

We kept at it. Not every repetition was a success, though. One time, I bumped the patient table as I reached in to give Callie her food. The slight jarring spooked her, and she backed out. Amazingly, she retreated only partway and came back when I held up the food.

After what seemed like an eternity, the scanner stopped.

Callie was waiting for me at the end of the patient table. I gave her a big hug and a handful of hot dog before escaping to the control room, where Robert was scrolling through a series of images on his screen. Most had snow. Occasionally something doglike would appear at the bottom of the screen, only to disappear a few images later.

Nothing.

"She wasn't in the field of view," Robert said.

My heart sank. Callie had performed so well, but her position during the localizer had been different. Without knowing it, we had programmed the scanner for the wrong location. Callie's brain was nowhere to be seen.

We had come too far to give up, and Callie showed me that she could do this.

"Let's switch to a dorsal orientation," I said.

Staring at Callie's localizer image, I realized what should have been obvious all along. The dog brain is longer front to back than it is from top to bottom. To better match the flattened shape of the brain, it made sense to take slices from top to bottom, which is called the *dorsal* orientation. Unless we took very thick slices or a lot of them, the FOV is a rectangular brick that is larger in the plane of the slices. By rotating the FOV to better match the flattened shape of Callie's head, we would be much more likely to capture her brain regardless of where she put her head down.

Using the cursor, Robert rotated the FOV ninety degrees. It was now aligned parallel to Callie's brain.

Between the scan and the fiddling at the console, we had burned up another thirty minutes. I broke my internal vow to switch to Mc-Kenzie. "Let's try this one more time. Then we'll give McKenzie a shot." Callie was lying on her side next to me. She was tired. I reached down to stroke her. Her tail thwacked the floor, indicating that she still had some juice left.

By now the team had settled into a routine. Rebeccah secured the earmuffs and Andrew took up his position at the rear of the scanner. Callie, now either bored or depleted of energy, sauntered into the magnet. She didn't so much as flinch when the sequence started.

Peas and hot dogs.

This time we blazed through the repetitions. After about ten of each type, I turned to Andrew.

"She's doing really well," I said. "Let's see if the new orientation worked." He nodded and gave the "cut" signal to Robert. Callie and I both ran to the control room. Everyone was already staring at the console.

Brains. We had brains.

As Robert scrolled through the sequence of images, you could plainly see cross sections through Callie's brain. We had captured

McKenzie wearing her wrapped-up earmuffs.
(Bryan Meltz)

sixty scans, and more than half of them contained an image of a brain.

I was elated. We had surpassed my hope of getting ten good images.

Everyone cheered and high-fived.

Lisa bent down to embrace Callie. "You did it!"

Callie licked her face.

Sinyeob just shook his head in disbelief, while Andrew summed it up: "Wow."

While everyone crowded around the computer, I sank into a chair, utterly exhausted. I hadn't realized how intense the last several hours had been. But now the adrenaline that had kept me going drained away, and I crashed. Same for Callie. She had already made her way to Melissa's pup tent for some quiet time and was sacked out.

But we weren't done. Now it was McKenzie's turn.

Callie had forged the way. Based on what we had learned about the noise and the earmuffs, we wouldn't have to waste any time with McKenzie.

Rebeccah worked her magic with the earmuffs. While Callie wore the small size, McKenzie had to wear mediums. Fully wrapped, McKenzie looked like she was wearing a turban.

Because every dog is a different size and shape, the scanner would again need to go through the shimming and localizer sequence for McKenzie. Melissa and Mark got her settled in the scanner and gave a thumbs-up to start the scan.

McKenzie reacted the same way Callie had. As soon as the buzzing started, she scooted out of the magnet. We did this three, four times, and despite the earmuffs, McKenzie was not having any part of it.

"What do you think we should do?" I asked Mark.

"The problem seems to be the sudden onset," he said. "The dogs are comfortable in the magnet when it's quiet, but the scanner starts without warning and scares them."

I turned to Sinyeob: "Can we start the scan before the dog is in there, and then get her to go in while it's running?"

He shook his head. "No, the scanner won't run with nothing inside."

"Maybe we could mask the transition," Mark suggested. "I'll make some noise before the scan starts to distract McKenzie."

Now on the fifth try, Melissa once again coaxed McKenzie into the MRI. Mark started whooping and hollering at her. Then the shimming began. Maybe it was Mark's carrying on, or maybe she had finally gotten used to it, but McKenzie stayed put. At least until the klaxon of the localizer began.

She'd been so close.

"Did it complete the shim?"

"Yes," Sinyeob said, "but she moved before the localizer." It was almost five o'clock, and we were just about out of time.

"Let's skip the localizer and go right to the functionals," I said.

"The chin rest should put her head in the same location as Callie's. We'll just use the same orientation and field of view that we used on Callie."

Andrew took up his position at the rear of the scanner and prepared to cue Melissa on peas and hot dogs. Mark started carrying on, making a ruckus to distract McKenzie from the sudden onset of the functional scan. Robert hit the start button.

I fully expected to see McKenzie's butt start backing out of the magnet. But she didn't. Mark stopped hollering. Images started appearing on the scanner console. At first, a nose poked into the field of view. Then part of a brain. And a little more. And then it would disappear, only to reemerge a few seconds later.

McKenzie was staying in the scanner. Melissa was putting up hand signals and feeding peas and hot dogs. From outward appearances, McKenzie was doing even better with this part of the scan than Callie had. The images popping up on the screen clearly showed McKenzie's brain, and they weren't moving, which meant that she was holding her head perfectly still.

The only problem was that her head was on the edge of the field of view. Even though she wasn't moving, we were capturing only the front half of her brain. This was a direct consequence of setting the field of view without a localizer image. We'd shot blind and missed by an inch.

I let the functional sequence run for the full five minutes. Even if we wouldn't be able to use her data this time, it was good training for Melissa and McKenzie. When it was done, I gave them the report.

"The good news is that McKenzie held her head still," I said. "The bad news is that we got only half her brain."

"McKenzie's not too tired," Melissa said. "We could try again."

"If you could get her to scoot her head forward an inch," I said, "that would help."

Mark, Melissa, I, Sinyeob, Robert, Lisa, and Kristina
study the first functional images.
(Bryan Meltz)

Once again, everyone took their positions, and with McKenzie re-settled in the magnet, we went through the protocol for what seemed like the hundredth time that day. This time, her head was closer to the center of the field of view. Some images were still clipped, but overall, the run looked very good.

Between Callie and McKenzie, we had exceeded our goal of acquiring a sequence of ten functional images. Even if we had only a partial scan of McKenzie, we had almost one hundred images of both dogs — enough to do a crude analysis of brain activation comparing hand signals for peas and hot dogs.

When we got home, Callie ran right to the kitchen. Even though she had spent the entire afternoon consuming peas and hot dogs, her appetite had not diminished in the slightest. She stood expectantly next to her food bowl. *So, where's my dinner?*

Lyra detected the foreign smells of the hospital and sniffed Callie from head to tail. Satisfied that it was still Callie, Lyra wagged her tail and let out a few yippy barks of recognition.

Helen plopped down on the sofa.

"So," I said, "what did you think?"

"It was pretty cool."

"Science doesn't always go the way they teach it at school," I said. "You never know what will happen." I paused and continued. "I'm really glad you came today. It was fun having you there."

Helen nodded her head, and I gave her a hug.

16

A New World

THAT NIGHT, CALLIE CURLED UP in her usual spot on the bed. Exhausted from the day of scanning, she immediately fell into a deep sleep, snoring softly. But I was still too jacked up; for the second night in a row, I didn't sleep. Images of Callie's brain danced in my head. But I had no idea what we had actually done. Dog brains were going to be our new world, but we had no map.

It turned out that the dogs' brains looked nothing like a human's brain. Apart from the size difference, many of the landmarks that I had become accustomed to seeing in human brains were either absent in the dogs or distorted into unrecognizable shapes. Now that we had dog brain scans, Andrew and I would have to grapple with interpreting all that data.

Dog brains and human brains differ in two important ways: structure and function. Brain structure refers to its shape. You don't need to be a neuroscientist to see that humans and dogs have brains of different shapes. The brain structure consists of the different parts of the brain and their location relative to one another. This is why neuroscientists refer to prominent parts of the brain as landmarks. Obvi-

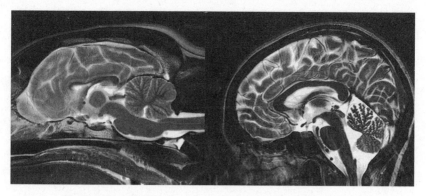

The dog brain (left) and the human brain (right). (Not to scale.)
(Dog brain image by permission of Thomas Fletcher, University of Minnesota;
human brain by Gregory Berns)

ous landmarks in all brains include the brainstem and spinal cord, the cerebellum, the ventricles (which make the cerebrospinal fluid), the corpus callosum (which connects the left and right sides of the brain), and a few structures in the basal ganglia — part of MacLean's reptilian brain.

But even with these landmarks, the largest part of the brain — the cerebral cortex — is radically different in dogs and humans. Presumably, that is what makes us different from each other.

Imagine comparing a map of the United States to a map of France. What can you deduce about these countries from looking at their maps? There is an obvious difference in size, but that doesn't say much. Based on the arrangement of roads, the maps would give you a sense of where hubs of activity lie. Many roads lead to Paris, and you might correctly conclude that Paris is a key center in France. You would also notice important port cities like Marseille and Bordeaux and guess that these cities are centers of trade. In comparison, there is no obvious center of activity in the United States, but the road map would give you much of the same kind of information. The Northeast Corridor, from Washington, DC, to New York, immediately stands out, and you would be correct in assuming that this region is a critical center of government and economic activity. Similarly, coastal cities like Boston, Houston, Los Angeles, and San Francisco would stand out as centers of trade.

And so it is with the brains of dogs and humans. Even though we don't know the exact function of different parts of the canine brain, we can make educated guesses based on what we have learned about other brains. Using landmarks that are common to both brains, we can begin to construct a more precise functional map of the dog brain.

But where to begin?

At first glance, dog brains didn't look much like human brains at all, so it wasn't apparent how much of the vast human neuroscience literature we could use. As I lay awake, I pictured the basic divisions of the human brain and tried to imagine how these might look in a dog's brain. It was very much like looking at a map of a foreign country.

If we think of the brain as a gigantic computer, information goes in, the brain does something with it, and an action is produced, often in the form of movement. In this manner, inputs and outputs form the first great divide in the brain.

Inputs are relatively easy to understand. All information that flows into our brains must come through the five senses: vision, hearing, touch, smell, and taste. From the scientist's point of view, inputs can be controlled during an experiment. For the experiment we had just accomplished, we controlled the visual channel by the hand signals we gave, and we controlled smell and taste by giving either peas or hot dogs.

Outputs are also easy to understand, especially if we consider movement as the main output of the brain. The earliest fMRI experiments had human subjects lying in the MRI and tapping their fingers for periods of thirty seconds. When the subjects tapped their fingers, activity in the part of the brain that controlled the hand was plainly visible.

The central sulcus is a groove in the human brain that runs almost vertically down the outside of each hemisphere. Everything behind the central sulcus is broadly concerned with inputs and everything in front with outputs. It is a defining landmark that divides the frontal lobe in front of the groove from the parietal lobe behind. The frontal bank of the central sulcus, it's important to note, contains the neurons that control movement of all the parts of the body. Toward the bottom of this groove, above the ear, we find neurons that control the hand and mouth, and as we move up toward the crown of the head, we find neurons that control the legs. The neurons found along the sulcus control the opposite side of the body. When you move your right hand, a portion of the left central sulcus will become active, and this can be seen easily with fMRI.

In contrast, the neurons behind the central sulcus respond when the corresponding parts of the body are touched. These are the primary sensory neurons. As you move farther toward the back of the head, the functions of the neurons become multimodal, meaning they integrate the inputs from many senses. At the very back of the head, we find the primary visual area, which receives inputs from the eyes.

Another obvious landmark of the human brain is the protuberance along the sides of the brain, just above the ear. This is the temporal lobe. Sitting directly next to the ear, parts of the temporal lobe are concerned with hearing. Other parts of the temporal lobe, along the inner crease next to the rest of the brain, contain structures critical for memory.

With the dog brain, the first thing you notice is that, apart from being smaller, it has a lot fewer folds. The massive amount of folding in the human brain is the solution that evolved to cram more brain into a small space. If you could flatten out the brain, you would find that all the neurons are contained in a thin sheet just a few millimeters thick. It's like taking a very large sheet of paper and crumpling it up into a ball. Once crumpled, a very large area can be made to fit in a small space, like the skull.

The different amount of folding in the dog brain means that the usual landmarks, like the central sulcus, don't exist. We can point to only the front and back of the brain and sort of make out the temporal lobe. The next thing you notice is that the dog doesn't seem to have much of a frontal lobe at all. This is the area that really distinguishes humans from other primates. Humans have the largest frontal lobes of any animal. Because the frontal lobes of the brain are mostly concerned with outputs — in other words, doing things — we think that this part of the brain expanded in humans to accommodate higher-order cognitive functions. Uniquely human functions that reside in the frontal lobe include language and the related ability to think symbolically; the ability to think abstractly about the future and past, which leads to planning; and the ability to mentalize what other people might be thinking.

Although the dog brain looks, at first glance, like a scaled-down version of the human brain, there is one area that is noticeably larger in the dog. The part of the brain concerned with smell, called the olfactory bulb, is huge in the dog brain. When the dog brain is viewed in the dorsal plane at the level of the eyes, the olfactory bulb looks like a rocket ship. There is no human equivalent of this part of the brain. The dog's olfactory bulb and the parts of the brain surrounding it compose almost a tenth of the total volume. Obviously, smell is important to dogs, but almost nothing is known about how this part of their brain works. That research would have to wait.

We had achieved the first milestone of success in the Dog Project by acquiring a sequence of functional images in both dogs. Over the next few days, we would match up the images with the timing data from the experiment. If everything worked, we would soon have a

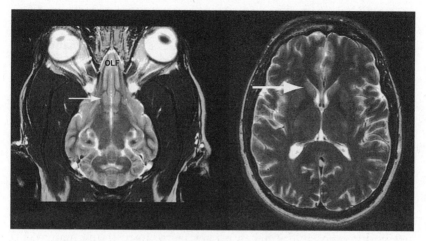

Dorsal plane view of the dog brain showing the olfactory bulb (left)
and the corresponding view of the human brain (right).
The arrows point to the caudate in both brains.
(Dog brain image by permission of Thomas Fletcher, University of
Minnesota; human brain by Gregory Berns)

picture of the dogs' brains that showed which parts responded to the
signals for peas and hot dogs.

But what would that tell us?

The whole of the Dog Project hinged on the promise of figuring
out what dogs think. Even if we succeeded in finding the parts of the
brain that responded to different hand signals, that wouldn't necessarily mean that we knew what the dogs were thinking. To answer
this deeper question, we would have to interpret the patterns of activation based on similar patterns in humans. If we saw activity in
parts of the dog brain that we could identify, and we knew what those
parts did in humans, we could begin to build a functional map of the
canine brain. Using the concept of *homology*, we could infer canine
thought processes from their human equivalents.

This was a shaky premise.

In recent years, there has been a bit of a scientific backlash against
neuroimaging. Functional MRI has made it easy to dream up poorly

controlled experiments and have groups of undergraduates go into the scanner. Many scientists, eager to get a quick publication in a high-profile journal, overinterpreted the patterns of activity they found in the human brain. It became commonplace to point to activity in a particular brain region and interpret that as evidence for a particular emotion or other cognitive function. It was too easy to observe activation of a structure and conclude, for example, that the person was feeling happy or sad or fearful or some other emotional state based on the scientist's assumptions of what different brain regions did. Eventually, neuroscientists termed this type of reasoning *reverse inference*, and it became a key factor in rejecting many fMRI papers.

I had always felt that the criticism of reverse inference, usually uttered with the same contempt one would have for a bag of doo-doo, was overblown. I wouldn't fault scientists for overinterpreting their data. If I doubted their conclusions, I could always look at their results and draw my own inferences. If I didn't believe their results, I wouldn't cite them in my papers. Good and valid conclusions stand the test of time, while false ones fade into obscurity and are eventually forgotten.

The Dog Project would not only be relying on reverse inference, it would depend on reverse inference of a dog's brain as if it were a human's. *Interspecies reverse inference.* I could already imagine what my colleagues would say about this.

Fortunately, Andrew and I had decided to stick with what we knew — the reward system. Our task of deciphering function in the dog brain was going to be a lot easier. Unlike the cortex, with its labyrinthine folds, the reward system belongs to the evolutionarily older reptilian part of the brain. The heart of the reward system is the caudate. Because it is so ancient, all mammals have a caudate, and lucky for us, it looks pretty much the same in dogs and humans.

While neuroscientists can quibble about reverse inference in the cortex, when we did an analysis of reverse inference in the caudate, we found that activity in this region is almost always associated with the expectation of something good. As long as we stuck to the caudate, we would be safe in interpreting activity in this part of the dog's brain as being a signal of a positive feeling. Everything else we found would have to be interpreted with caution.

Even if we limited ourselves to simple questions of whether the dog had positive feelings based on caudate activation, we could still accomplish a lot with brain imaging. No longer would we be stuck interpreting dogs' behavior based on tail wagging, which is an imperfect indicator of the emotional state of a dog. Dogs wag their tails when they're happy, when they're anxious, or when they're unsure of what else to do. I still wanted to know if our dogs reciprocated our love for them in any way. And although love is a complicated human emotion, the positive aspects of it have been consistently associated with caudate activation.

The first experiment was a proof of concept. Before we could move on to complicated questions, like love, we first had to demonstrate that we could measure caudate activity in the dog. But that wouldn't be enough. We would have to show that we could interpret that activity in terms of how much the dogs liked something. Because hot dogs are so much better than peas, especially to a dog, the hand signal for hot dogs should cause more caudate activity than the signal for peas.

It seemed simple. But like everything else about the Dog Project, it was also completely wrong.

17

Peas and Hot Dogs

WITH THE APPARENT SUCCESS of the first scan session, Andrew quickly set to analyzing the data. We were giddy that we had not only captured images of the dogs' brains, but that we had also succeeded in getting several runs of functional scans. These functional runs ranged in length from two to five minutes. At first glance, it looked like we had far exceeded our goal of acquiring a sequence of ten images. In McKenzie's case, we had one run of 120 images. However, it soon became apparent that figuring out what we had actually captured was going to be far more difficult than we had imagined.

Once the excitement of looking at dog brains began to fade, the first thing we noticed was that the dogs didn't keep their heads in exactly the same position. There were stretches of about ten seconds where the images appeared steady, almost as good as a scan of a human. And then the dog would move out of the field of view. This would be followed a few seconds later by the head reappearing, but not in exactly the same spot.

It was during these gaps that we had handed the treats to the dogs. Normally, a human would be lying on his back, nose up, almost touching the inside of the head coil. But because the dogs were in a sphinx position, they were facing toward the far end of the scanner, where Melissa and I were giving hand signals and dispensing the treats. At the end of each hand signal, we would grab either a pea or a tiny cube of hot dog and reach all the way to the dogs to let them eat it from our fingertips. Of course, there was no way the dogs could keep their heads still while eating, but they had seemed to settle down pretty quickly. Looking at the MRI images, it became apparent that the inconsistency of positioning was a bigger problem than we had expected.

Somehow, we needed to figure out a way to compensate for the different head positions. In the terminology of fMRI data processing, this is called *motion correction*. Normally, motion correction is done digitally with special computer software after all the data are collected. The software can figure this out automatically by shifting each image until it exactly overlays the first one of the sequence. For humans, it is pretty simple because they don't move much, and the corrections are generally less than a few millimeters. Because the dogs didn't return to the same position each time, the brain had shifted in location too much for the automated software to find it.

Instead, we reverted to an old-school approach of digitally defining landmarks in the brain. First, we identified blocks of scans in which the dog's head was in a steady position, regardless of where it was in the field of view. For each of these blocks, we then placed four digital markers on identifiable landmarks: the olfactory bulb at the front of the brain, the left and right sides of the brain, and the brainstem at the bottom. Then we used software to shift the images until the landmarks were all aligned. The movement can be described by how far you slide it, which is called *translation*, and by how much it rotates. If the dog moved its head to the left, we digitally shifted it back to the right to keep it centered. If she pitched her nose up a little, we digitally rotated the image so her nose was level.

Amazingly, this worked. When we viewed the sequence of images in a rapid movie loop, the head now appeared to remain steady in one position. Even Callie, who was not as consistent as McKenzie, appeared stable in the motion-corrected images. We were ready to analyze the actual activation patterns.

Naturally, we assumed that a hand signal indicating hot dog would be much more exciting than one for peas and that this difference would be reflected in the dogs' brains. To decode how their brains processed these hand signals, we needed to compare the brain responses for each dog to the hot dog and pea signals. Using a standard technique in brain imaging, we separated all the trials into groups of hot dogs or peas. Next, we calculated the average brain response to each of these signals and subtracted the average pea response from the average hot dog response. If our hypothesis was correct, the difference would show up in the parts of the brain that respond to reward.

Instead, we got nothing. No matter how many different ways we looked at the brain responses, it didn't appear that the dogs distinguished between the hand signals at all.

Melissa had said from the beginning of the Dog Project that McKenzie preferred toys to food. But we couldn't give her toys to play with in the scanner. Think of the head movement that would cause as she shook her head back and forth! There wasn't any way around using food as the reward. Callie, of course, was highly food motivated. In fact, she might have loved food too much.

Callie's food drive was a key factor in her learning the task so quickly. Although she still looked like a tightly wound spring, ready to uncoil in a burst of energy, the prospect of a hot dog could keep her still, at least for a minute or so. There was no doubt that she loved hot dogs, and I saw no reason to use anything else during training.

It didn't seem to matter what brand of hot dogs I used. Kosher beef dogs seemed like a natural place to start, but then we started expanding her palate. We tried turkey dogs. One brand had a deep, smoky aroma, and this seemed particularly effective. It was so infused with smoke, in fact, that no amount of washing could remove the smell from my hands. But Callie really liked it. She could hear that particular package being opened from the other side of the house, and before the hot dog was fully removed from its plastic bag, she was there at my feet, wagging her whole rump, sweeping the floor with her skinny rat-tail. With that reaction, it was hard to imagine anything better for training.

But then again, Callie was an inveterate poo eater. Did she really prefer hot dogs to peas? What if she was completely indiscriminate and ate everything?

This was potentially a big problem. If the dogs didn't care whether they ate hot dogs or peas, then the hand signals would be meaningless. They knew that they would get a treat for putting their head in the rest, so if it didn't matter which treat, there would be no motivation to pay attention to the hand signals. We probably should have dealt with this before the first scan session, but science is imperfect, and you can't predict how experiments will go.

Before we went any further in the Dog Project, I decided it would be worth testing Callie's discrimination between hot dogs and peas. If Callie were a human, it would be a simple matter to ask her which one she liked better. Since she couldn't speak, I was stuck with the classic problem of guessing what was in her mind by observing her behavior. The trick was to devise a series of tests that would force her to reveal whether she preferred hot dogs or peas.

My first idea was to give Callie a choice between hot dogs and peas. Thinking like a human, I reasoned that if I placed a hot dog and a pea on a plate, the one she ate first would have to be her favorite.

This was a two-person operation. Every time I opened a bag of food, Callie was right there at my feet, where she remained glued until I gave her what she wanted. I had to enlist Kat's help to hold her off while I prepared the test.

While Kat held Callie on one side of the living room, I carefully placed a pea and a piece of hot dog on a plate at the opposite end of the room.

"Go!" I exclaimed.

Kat released her and Callie darted to the plate. Without any hesitation, she licked up the hot dog first and then the pea. So far, so good.

"Hot dog!" I called out.

Feeling very pleased with my ingenuity, we set up to try again, except this time I reversed the location of the hot dog and pea. I nodded to Kat, and she released Callie. Once again, she made a beeline to the plate.

And lapped up the pea.

Okay, I thought, *we can't expect perfection.* Maybe she was just excited.

"She ate the pea," I called out. "Let's try again." Kat rolled her eyes, but humored me anyway. We repeated this ten times, and every single time Callie went to the left side of the plate, which is where the hot dog was located on the first trial.

"Maybe she knows that she will get both treats, and that's why she goes to the same side," Kat said.

Yes, of course. *Must think like a dog.* If I were Callie, I would just scoop up whatever was closest and move on to the next one since I would be getting both anyway.

"What if I picked up the treat she doesn't choose first?" I said. "That way, she will have to make a choice."

We reloaded the plate, and Kat let her go. Just like before, Callie lapped up the pea on the left side of the plate. This time, I snatched the hot dog away just as she started to make a move toward it.

If a dog could shrug its shoulders, Callie surely would have. She trotted back to Kat and waited for another round. But in the end it didn't seem to make any difference. Callie just kept going to the left side, where the pea was placed. How could she prefer a pea to a hot dog? The hot dog was loaded with carnivorous goodness, perfectly suited to her Paleolithic instincts.

After about ten trials with the pea on the left, she finally paused and noticed the hot dog on the right side. In fact, this was the first time I had ever seen her pause with food in front of her nose. As if to say, *Hey, where did this come from?* she went for it.

And then she was stuck on the right.

Hot dog or pea, it didn't make a difference. No matter how many times I placed a pea on the right, she wouldn't track the hot dog.

Kat shook her head and said, "Do you still need me?"

"No," I said. "I have another idea."

I looked at Lyra, who had been watching the experiment from the sidelines. She had long gotten used to seeing Callie work for treats. If she waited long enough, she too would get to partake in the spoils, just for looking pretty. She perked up when I turned to her.

"Lyra, come here, sweetie!"

The three of us — Callie, Lyra, and I — padded into the kitchen.

Now, with the plate on the counter, I placed a pea on the left side of the plate and a hot dog on the right. Both dogs were rapt with anticipation. Quickly, I placed the plate on the floor.

As expected, Callie lurched for the right side, where she was still fixated. Before she could nab the hot dog, I grabbed it away and fed it to Lyra. A momentary look of confusion flashed across Callie's face. Lyra was delighted and started to drool.

Surely, I thought, this would make Callie think about her choice.

It didn't. Even now, with Callie's usual avarice, she continued to perseverate on one side of the plate. Either she really didn't care about the difference between hot dogs and peas, or her brain was executing a simple rule: stick with the same side as long as it has something good.

The next day, I asked Mark about Callie's tendency to stick to one side.

"That's common," he said. "Some dogs are naturally right- or left-sided. Other dogs will remain with whatever attracted them first. Some will remain with wherever they were rewarded last. Other dogs will relax and cognitively evaluate each situation or wait for a cue."

In fact, the development of a side preference was documented in a series of cognitive experiments in dogs in 2007. Researchers at the University of Michigan were attempting to determine if dogs had a concept of quantity. Does a dog know that two pieces of food are better than one? It seems obvious to us humans, but if you think about it, "quantity" is really quite an advanced concept. It requires some knowledge of the physics of the world, that larger volumes hold more stuff, and that more is better. Although there is some evidence that infants can discriminate basic differences between, say, one and two objects, the cognitive skill called *numeracy* doesn't fully develop in humans until early childhood.

The researchers wanted to know if dogs had abilities similar to human infants. They tested twenty-nine dogs on a task very similar to what I had concocted in my kitchen. Plates with different amounts of food were offered to the dogs, and it was observed which plate the dogs chose. Most of the dogs chose the plate with more food, although not all the time. It was not clear whether the dogs actually had a sense of quantity or whether they were responding to perceptual cues of bigger piles of food. Either way, it was also noted that eight of the dogs had to be excluded from the analysis because they developed a side preference regardless of the quantity presented.

• • •

So Callie had a strong tendency for side preference, which seems to be a normal variant among dogs. Still, I was disappointed that my feisty rescue wasn't an Einstein. I had no idea about McKenzie's preferences, but since at least one of our subjects couldn't tell us the difference between peas and hot dogs, a change in the experiment was in order.

The next day, I reported my findings to the lab.

Andrew was disappointed. "If they don't care about the difference between peas and hot dogs, how will our experiment work?"

"It won't," I replied. "If peas and hot dogs are the same to the dogs, then the hand signals convey no useful information. As long as they put their head on the rest, they know they will get a treat. They don't care which."

Nobody could understand why the dogs didn't discriminate between the two foods. We were all stuck thinking like humans. We had to think like a dog.

"What if we just get rid of the peas?" I mused.

"You mean a reward versus no-reward experiment?" Andrew asked.

"Exactly. Even if the dogs don't care about hot dogs versus peas, surely they care about hot dogs versus nothing."

Andrew nodded in agreement.

It wouldn't even require any new training. We already had the two hand signals. Left hand up meant "hot dog." Now, two hands pointing toward each other would mean "no hot dog" instead of "pea."

"Don't you think the dogs will get irritated and stop doing the task?" Andrew asked.

It was a good question. If it were me in the MRI simulator, I would scoot right out of there as soon as I realized I wasn't getting food all the time. Psychologists call this *extinction*, which means that if you stop rewarding a previously learned behavior, the behavior will eventually stop.

But dogs might see it differently. Not rewarding every trial might increase their motivation. This is called *variable reinforcement* — VR for short. VR is very common in animal experiments. A VR10 schedule means that sometimes the subject is rewarded but, on average, only once every ten trials. The unpredictability of VR tends to make animals more attentive and work harder to obtain the reward.

Something as drastic as VR10 would not work in our experiment. I just couldn't see Callie sitting still for ten repetitions of hand signals to get just a tiny cube of hot dog. If I were in her position, I would begin to wonder after the third repetition without a treat. By about the fifth repetition without food, I would probably give up entirely and quit the experiment. I suspected Callie wouldn't put up with it either. More important, it would result in an imbalance in the number of observations collected in the scanner. If we had ten trials of the no-reward hand signal to every one of the reward hand signal, it wouldn't be an even comparison. We needed an equal number of rewarded and unrewarded trials for this to work. The solution was simple: VR2.

A VR2 schedule means that roughly half the trials will be rewarded (two trials for every one that is rewarded). This would give an equal number of observations for both hand signals. As long as we didn't simply alternate, which would make it completely predictable for the dogs, then they should stay highly motivated.

That evening, I tried VR2 on Callie.

As usual, the rustling of the hot dog bag called her to the kitchen.

"Wanna do some training?" I said in my high-pitched doggie voice.

Callie cocked her head and tore off into the living room. When I got there, she was already in the tube with her head in the chin rest. To warm up, we went through several trials as usual. Left hand up, hold it for ten seconds, and then reward. When she seemed settled in, I flipped the two-hand signal that had previously meant peas. This time, after ten seconds, instead of giving her a pea, I just touched her forehead. She thought a pea was coming and tried to lick my hand. With nothing there, she looked puzzled.

I pointed to the chin rest and said, "Touch."

Callie quickly placed her head down. To make sure she wasn't confused, I immediately showed her the reward hand signal, and rather than wait ten seconds, rewarded her right away. The next trial, I gave the two-handed, no-reward signal and quickly ended the trial with a touch on the head. We repeated this for about five minutes, and amazingly, she didn't get bored or leave the simulator. Instead, her posture and attentiveness improved. Her head positioning became more consistent, and her eyes were fixed in attention on my hands. Now when I showed the reward signal, I could see her pupils dilate, indicating a high level of positive arousal. And she remained motionless.

VR2 was a success! If Callie could catch on so quickly, I was sure McKenzie would too. And with her pupils dilating, it was clear that Callie now cared about the hand signals.

If this didn't work, nothing would. We were ready.

18

Through a Dog's Eyes

WE DIDN'T HAVE MUCH TIME to get Callie and McKenzie trained on the new version of the task. We could have taken longer with the training, but the logistics of finding a day when Mark, Melissa, Rebeccah, and the scanner were all available dictated the schedule, and the next available time that everyone could meet again was only two weeks away. If we missed the window in two weeks, we would have to wait another month to book a large chunk of time at the scanner. The pressure was on.

At least we knew the dogs could do this. Each time we had gone to the scanner, we had accomplished more than I had expected, and I was counting on this next time to be no different. The dogs knew what they had to do. The real uncertainty was how much data we would be able to collect and whether it would be enough to demonstrate caudate activity.

The fMRI signal is very weak. We measure activity as the relative change in signal intensity from some baseline level. But even in the best of circumstances the signal intensity rises by less than 1 percent. To make matters worse, the fMRI signal is noisy. The noise, which comes from heart rate variability, breathing, and even the electronics of the scanner, causes fluctuations in the signal that are ten times as much as the thing we are looking for. The signal-to-noise ratio (SNR) of fMRI is therefore quite low. Fortunately, the noise is random. If we collected enough repetitions during the experiment, we could average the fMRI signals from each, and the effects of noise would be diminished.

Often, when doing an experiment for the first time, you don't know how big the signal is, so you have to make an educated guess at how many repetitions will be required to detect it. The Dog Project was on the verge of moving beyond a cute dog trick and into the realm of legitimate science. But to make this jump, we would first need to figure out how many repetitions would be required.

Andrew and I took a close look at what we had collected on the first scan day. Even though we had failed to find any differences between peas and hot dogs, there was still useful information in the data. We could estimate the SNR of the dog brain and, from that, determine how many repetitions Callie and McKenzie would have to do at the next scan.

Andrew zoomed in on the caudate of McKenzie's brain. He pulled up a graph of the level of activity in the caudate for each scan that we had acquired. The first few scans had no signal because McKenzie hadn't placed her head in the head coil until about the twentieth scan. But then it looked like noise. It was hard to tell how much of the noise was because of the usual sources or her moving slightly during the scan. The size of the fluctuations measured about 15 percent of the overall signal. This was much higher than in human studies.

I let out a sigh and said, "We would need a thousand repetitions to get the SNR up to a reasonable level." Neither dog nor human would sit still for that long. "That has to be from movement."

"It is," Andrew said. "Check this out." Andrew scrolled through the sequence of McKenzie's images. This had the effect of creating a movie of her brain. It compressed the five-minute scan session into thirty seconds. Even though we had captured only half of her brain, the movie made clear that although McKenzie had been in the head coil for the whole session, she was still moving. Not much. But just enough to cause artifacts. Callie's brain movie looked similar.

In fairness to the dogs, they had done what we asked of them. The degree of movement we were talking about was a matter of millimeters. During the stress of the scan session, neither Melissa nor I noticed it. Not that we could have done much about it on the fly.

"Well," I said, "we have two weeks to train them not to move."

Andrew looked skeptical.

The dogs would have to move less than two millimeters, but they would have to hold still only during the period from putting their head on the chin rest through the duration of the hand signal. After they got their hot dog, they could take their time swallowing and getting resettled in the chin rest. We needed enough time only for the fMRI signal to reach a peak and begin to decay — roughly ten to fifteen seconds. If the dogs remained motionless for that length of time, we calculated that twenty repetitions might be enough to get the SNR up to a usable level. That still left the structural scan, which we hadn't been able to obtain on either Callie or McKenzie. That scan would require the dogs to stay motionless for thirty seconds.

Thirty became the magic number. During training, we would have to gradually lengthen the time between the hand signal and the reward until the dogs could hold absolutely still for half a minute. If they could do that, we would be able to get the structural scan and plenty of functional repetitions to boot.

Callie didn't seem to mind the change in training procedures. At first, I felt a twinge of guilt every time I put up the signal for "no hot dog." Callie would stare at me impassively from her position in the mock head coil. *I'm in the head coil, why no hot dog?* Sometimes I would touch her lightly on the top of her head, indicating that I wanted her to try again. But this soon became superfluous.

It seemed cruel to withhold rewards, but I trusted Mark's advice and stuck to the VR2 training schedule.

Mark was right. After switching to variable reinforcement, Callie really started paying attention. She had no choice. With peas and hot dogs, she got rewarded every repetition, so there was no need to pay attention to what I was doing. Now she noticed every little movement. If my shoulder twitched, Callie's eyes darted to the side. It was so fast that had we not been staring directly at each other I would never have noticed.

I continued recording our training sessions. With a digital camera on a tripod, I could shoot directly over my left shoulder. Even though Callie and I were staring directly at each other during training, the camera picked up things that I hadn't noticed in real time. Mark and I reviewed these videos like football coaches on the Monday after game day. He critiqued my technique as we tried to eliminate all of my "tells." We wanted the dogs to be focused solely on the hand signals. Callie wasn't the only one who would have to hold perfectly still. So would I. Except for the hand signals, we didn't want any extraneous body movements.

We amped up the noise training too. Both Callie and McKenzie had reacted negatively to the sudden onset of the shimming and localizer sequences, so we incorporated recordings of those noises into the daily training as well. The more the dogs became accustomed to the sounds, the more comfortable they would be.

We even tried to make positive associations with the noise. I would start playing the scanner noise through the PA and call Callie to the living room. We would wrestle and play tug-of-war while the noise blasted away. Lyra would join in too. It took only a few days before Callie would run to the living room as soon as she heard the scanner noise playing. I would slip on the earmuffs and crank the amp to 95 decibels to give the full effect. She didn't care. Callie would just trot up the steps into the tube and plop down in the head coil, licking her lips and waiting for hot dogs.

It was during this intense training period that I think our relationship began to change. Rather than master-dog, or dominant-subordinate, we became a team. We were like pitcher and catcher. For lack of a better word, it was intimate.

There is something deeply personal about staring directly into another's eyes. Humans' eyes are unique. We have more white in our eyes than any other animal, which means that we can tell with extraordinary precision where other people are looking. One theory says that humans' eyes evolved this way as a means of nonverbal communication. Using nothing but eye movements, we can, for example, communicate to other people where they should direct their attention. Just as important, we can deduce a great deal about someone's thought processes just by observing where they are looking. Gazing directly at you? They are definitely interested. Gaze averted or roaming? Not so much.

Under normal circumstances, when I had looked into the eyes of animals, even our beloved pets, I never felt a strong reciprocal connection. Sure, they looked back, but the gulf between species was too great. It was like staring into an abyss with no clue as to what lurked behind those big brown eyes.

Now, eyeball-to-eyeball, I could see my reflection in Callie's eyes. Yes, she wanted hot dogs, but there was something more. Callie had been communicating with me the whole time. I had been the one who was blind to it. But now that we were staring at each other for minutes on end, there was no ignoring it. Subtleties of expression—

how she held her eyebrows, the tension in her ears, the drape of her lips, and, of course, where she directed her eyes — spoke volumes.

Now too late, I realized that Newton had done the same.

As dog trainers have known for a century, dogs are exquisitely sensitive to picking up cues in their environment. Dogs act with a *theory of behavior*, which is the broad scientific term for saying that dogs learn that certain behaviors lead to certain outcomes. This is the foundation of positive reinforcement. But staring into Callie's eyes, and watching how she stared back, I began to suspect that she was doing something more. She was noticing where my attention went.

The ability of dogs to track others' attention has only recently been appreciated. In 2004, researchers in Hungary tested the extent to which dogs used attentional cues from humans. They set up a series of experiments that included different types of fetching tasks that varied the face and body positions of the humans. The researchers wanted to know how dogs reacted to a human when they either faced each other or faced away and whether the visibility of the human's eyes made a difference. To hide the human's eyes from the dogs, the person was blindfolded. The researchers found that dogs were sensitive to the human's attention, but that it depended on the specific context. In tasks that were playlike, the dogs didn't seem to care whether the human was looking at them, but if the human commanded the task, then the dogs paid close attention to where the human was looking.

The evidence continues to accumulate that not only are dogs sensitive to where humans' attention is directed, but dogs are also sensitive to the social context. They know when it is appropriate to attend to their human's attention and when it is not. This means that dogs have more than a *theory of behavior*. They have a *theory of mind*.

In humans, theory of mind, or ToM, means that we can imagine what another person might be thinking. Reflecting the importance of humans' social lives, most of our large frontal lobes seem to be concerned with this function. We spend huge amounts of mental energy navigating the complex social structure of human society. Knowing how to read people and how to behave in distinct social settings is the difference between success and failure. And at the extreme, autism may represent a failure of the ToM system in the brain.

If dogs have ToM abilities, they are probably simpler than ours. The small frontal lobes in the dogs' brains are clear evidence of that. But even if dogs have only a rudimentary ToM, that would mean dogs are not just Pavlovian stimulus-response machines. It would mean that dogs might have about the same level of consciousness as a young child.

As Callie and I honed our performance, I had a growing sense that we were beginning to read each other's mind. Of course, there was no way to prove this. The thought was so crazy I didn't even voice it in the lab. But we were about to discover that my intuition wasn't completely off the mark.

The second scan day arrived on a drizzly February afternoon. Under cover of umbrellas, the entourage once again made the trek from lab to hospital. The novelty had faded somewhat, so fewer people were in attendance, and the overall atmosphere was calmer and more businesslike. Robert and Sinyeob greeted us at the scanner. This time they weren't laughing. Everyone knew the dogs could do this, and we were there to do science.

There was no need to fiddle with the scanner settings. Robert simply pulled up the final parameters from last time, and we were good to go. The plan was to do the shimming and localizer, two five-minute functional runs of hot dogs versus no hot dogs, and then a thirty-second structural scan. If there were no hiccups, we could blaze through the procedure in thirty minutes for each dog.

It really helped that everyone knew what to do now. Rebeccah worked her magic touch with the earmuffs and head wrap on Callie. Andrew took up his position at the rear of the magnet, ready to re-cord the repetition type — hot dog or no hot dog. Melissa and Mark settled McKenzie in her pup tent until it was her turn. I motioned to Callie to go into the scanner.

To avoid startling the dogs with the sudden onset of buzzing, Mark had hit on the great idea of playing the recordings that we had used during the training procedure. Every MRI has an intercom to allow communication between the patient and the technician. After Callie got settled in the chin rest, the team in the control room held an MP3 player up to the intercom and began playing the recording

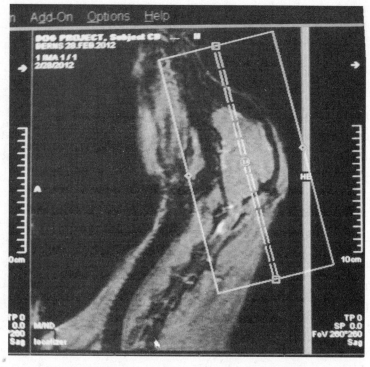

Callie's localizer with box indicating field of view.
(Gregory Berns)

of the localizer noise. Softly at first, then they gradually cranked up the volume. Pretty soon I could hear the familiar swarm of bees emanating from the speakers built into the magnet. Because it came on gradually, Callie didn't budge.

I nodded to Andrew. The buzzing continued. And then it stopped.

"What happened?" I asked. Andrew shrugged. Callie followed me into the control room. "Why did you stop the scan?"

Robert looked confused. "We didn't," he said. "Look."

There, on the computer screen, was a perfect image of Callie in profile. The image was sliced right down the middle of her head, giving a beautiful view of her brain and spinal cord. Robert had already placed the bounding box for the field of view in place. The use of the recording through the intercom had worked so well that neither Callie nor I had noticed when the real scan started!

With the field of view set, we cued up the functional runs. I showed Callie the container of hot dogs, and her eyes widened. All I had to do was point to the magnet, and she scooted in.

This time, we played the recordings from the functional sequence through the intercom. The volume was gradually increased, and after a few seconds, I could hear the real scans begin. They sounded almost identical. Callie didn't care. Her eyes were lasers on mine. I held up my left hand to indicate that she had done well and gave her a piece of hot dog.

We were off and running. I alternated repetitions for hot dogs and no hot dogs but kept it somewhat unpredictable, throwing in runs of two or three of the same trial type. Callie stayed cool as a cucumber. Every time I put up the sign for "no hot dog," she stared at me and waited until I put up the sign for "hot dog." I began to appreciate that rather than being disappointed during the no hot dog trials, Callie viewed those hand signals as uninformative. Being told that she wouldn't get a hot dog said nothing about when she would get one. This interpretation would soon be borne out by her brain activation.

Unlike the previous scan session, this time we were much more efficient. In short order, we had acquired two five-minute runs of functional scans, nearly four hundred images in total. The only thing that remained was the thirty-second structural scan. At this point, Callie looked either tired or bored, but in she went for the fourth time. The recordings of the structural sequence were slowly ramped up through the intercom, and then the real scan started. The structural sequence sounds much like the localizer, but Callie stayed put throughout.

She had done it. She hadn't moved at all. I ran around the scanner and gave her a handful of hot dog.

"You are such a good girl!" I exclaimed. "You are a SuperFeist!"

Robert already had the structurals on the screen. There, in breathtaking clarity, was the first detailed structural image of a completely awake dog. My jaw dropped. We had just acquired nearly four hundred functional scans and a structural image that rivaled anything we got in humans.

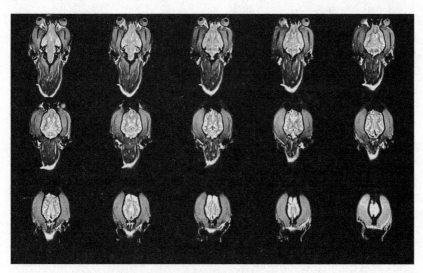

The first detailed structural image of
Callie's brain rivals the quality of human scans.
(*Gregory Berns*)

Even if McKenzie bombed on her turn, I was confident that we had achieved our goal of getting enough functional scans.

"How many repetitions did we get?"

"It looks like she did twenty trials with hot dog and nineteen trials of no hot dog," Andrew said.

"Damn," I marveled. "That should be plenty for analysis. Let's hope the SNR is high enough."

Callie sat down next to me. I looked into her eyes, and she knew. *Yeah, I'm the top dog.*

If I'd had any lingering concerns about Melissa and McKenzie, those quickly disappeared. The trick of playing the recordings through the intercom worked wonders for them too. We finally got a localizer image for McKenzie, which allowed us to precisely place the field of view to avoid chopping off half her brain this time. For the functional scans, Melissa was more collected than I had been. She really took her time with the repetitions, requiring McKenzie to hold still for fifteen seconds for each trial, where I had required Callie to hold still for only ten.

McKenzie was like a rock. Robert and I watched her images stream on the console in real time. She was not moving. Not at all. They blazed through the two functional runs, and, for the first time, we got a structural image of McKenzie's brain.

Two for two.

Not only was the day a complete success, but we had accomplished all of this in two hours — half the time of the previous session.

It was still exhausting. When you're locked face-to-face inside the magnet with jackhammers all around you, the level of concentration, for both dogs and humans, is intense. When Callie and I finally got home, we crashed together on the couch. We looked at each other once and then closed our eyes.

19

Eureka!

ANDREW DIDN'T WASTE ANY TIME. The next day, he had already begun the analysis of Callie's and McKenzie's data. Just like the peas and hot dogs experiment, the first and trickiest part of the analysis would be the motion correction. We had to carefully identify which scans contained brains and discard the ones in which the dogs moved too much. Animating the sequence of images in rapid speed helped make the task easier.

Andrew showed me the animation.

"Check this out," he said. A pixelated image of a dog's brain danced on the computer screen. For stretches of several frames, which were actually tens of seconds in real time, the image didn't move. Except the eyeballs, which darted left and right.

"This is Callie," Andrew continued. "She did really well. If we throw out the scans with movement artifacts, we still have 62 percent left for analysis." My heart swelled in pride at my beloved feist.

"That's amazing," I said. "That is five times better than the previous session. How about McKenzie?"

"Almost as good. We can keep 58 percent. She had sixteen hot dog trials and eleven no hot dog trials."

"Melissa was really making her hold still for a long time," I said.

"Yes," Andrew said, "but that means we'll have a lot of scans for each repetition."

We spent the next two days checking and rechecking each step of the analysis. To make sure that we didn't mistakenly confuse brain activation with movement artifacts, we kept ratcheting up our criteria for whether to keep a scan in the analysis. Andrew and I would stare at the animations, looking for even the slightest twitch of the head. Most of the head motion occurred when we gave the dogs hot dogs. This was no surprise. But we weren't interested in the brain response to hot dogs. We were interested in the response to the hand signals. When we were satisfied that we had identified and discarded all the scans with motion, the remaining scans showed that the dogs had held their heads with less than one millimeter of movement during the critical period of the hand signals. That was as good as humans do in the scanner. We were ready for the final step: comparing the activation between the two hand signals.

All fMRI experiments measure relative changes in brain activity between different conditions. With only two conditions — the signal for "hot dog" and the signal for "no hot dog" — all we had to do was subtract the brain activity in one condition from the other. The difference would show us which parts of the dogs' brains processed the meaning of the signals.

To do this, we usually calculate the difference in activity at every location in the brain and perform a statistical test to determine whether the results are real or simply random fluctuations in the fMRI signal. We then create a map from this analysis and overlay it on the structural image. By convention, neuroscientists use a color scheme that ranges from yellow for weak activations to bright red for strong ones. Andrew did the subtraction for Callie.

Everyone in the lab had been waiting for this moment and gathered around the computer screen.

There, overlaid on Callie's pyramidal-shaped brain, were several hot spots of yellow and red. We still didn't know what most of the brain was doing. It was important to stay focused on the one region that we knew a lot about.

"Zoom in on the caudate," I said.

Sure enough, an orangish blob sat squarely on top of the right caudate. There was no doubt. The lab stared in amazement and let out a collective gasp.

McKenzie's activation map was even stronger. Both dogs showed unmistakable proof of caudate activation to the signal for "hot dogs" but not the signal for "no hot dogs."

Had only one of the dogs shown caudate activation, it would be easy to dismiss as a fluke. But we were looking at caudate activation in both dogs. The odds of that happening by chance we calculated to be 1 in 100.

"Caudate activation in both dogs?" I said. "That it is no accident. That is real."

At dinner that night, I broke the good news to the girls.

"The Dog Project worked," I announced.

"What do you mean?" Kat asked.

"We found reward-system activation in both dogs."

"So," Kat said skeptically, "you discovered that dogs like hot dogs?"

"No," I replied. "We discovered that they understand the meaning of hand signals."

This was a crucial distinction. In fact, Andrew and I did observe caudate activation to the hot dogs. But because the delivery of the hot dogs also caused Callie and McKenzie to move their heads as they swallowed and licked their lips, we had to discard a high proportion of those scans. Even so, the caudate activation was still plainly evident. But for the reasons Kat implied, such a finding would not be very surprising. Everyone knows that dogs like food.

No, the big result was caudate activation to the hand signal for "hot dog" but not "no hot dog."

The Pavlovian behaviorists would say, "Ah, the dogs learned the association between a neutral stimulus — the hand signal — and an unconditioned response from the food. Nothing in the brain implies an understanding of meaning." Had we done the experiment like Pavlov, using the ringing of a bell, for example, or the turning on of a light in place of the hand signal, this would certainly be true. But we used hand gestures. Humans take it for granted that hand gestures convey a great deal of information, almost as much as the eyes. Is it possible that dogs place as much importance on hand movements as we do?

A growing body of evidence suggests that they do.

Brian Hare, an anthropologist at Duke University, has pioneered the study of social cognition in dogs, especially the extent to which they understand human social signals. In his initial experiments, Hare hid food in one of several possible locations in a room. A human would stand in the room and point to the correct location. When a dog entered the room, it was able to use the pointing cue to more quickly find the food. Often, the dogs did this on the first try, indicating that simple associative learning, like the behaviorists believed, could not explain dogs' ability to intuit the meaning of human social signals. Dogs seem to be particularly skilled at reading human signals. Hare later tested wolves and chimpanzees, and neither did as well as dogs.

Even at the dinner table, I could see that Callie was exquisitely attuned to our human social interaction. Lyra — not so much. But Callie sat in relaxed attention. Her head would swivel to whoever was speaking. Although she couldn't understand all the words, if someone said one of the words she did know, like *walk*, she would run to that person and start wagging her tail vigorously. More than speech, I knew she understood hand signals, because all I had to do was point at the MRI tube, and she would go in.

Now we were faced with the conundrum of reverse inference.

If we had been studying humans, the interpretation of the caudate activation would be pretty simple. In fact my colleagues and I had done exactly this kind of experiment ten years earlier. Instead of hot dogs, we used Kool-Aid. In that experiment, our human subjects lay in the scanner with a tube snaking into their mouths. When a green light appeared on a computer screen, the subjects would have to press a button, and then, a few seconds later, they would get a squirt of Kool-Aid on their tongues. Just like Callie and McKenzie, the humans' caudates activated to the signal indicating impending Kool-Aid. Since then, this result has been replicated dozens of times by us and other researchers. The advantage with humans, of course, is you can ask them what they thought and felt in response to the signals.

Inevitably, people attributed meaning to the signals. For some people, signals set up a state of anticipation. Indeed, a state of heightened anticipation, especially of something good, is probably the most universally experienced emotion associated with caudate activation. This state of anticipation drives people to get what they desire. In the extreme, we call it *craving*, and dysfunctional caudate activity is generally believed to be associated with addictions. Now, if simple computer cues are replaced with more humanlike cues, then caudate activity is even greater. For humans, there appears to be a bonus effect in the caudate to social cues, even if they convey the same information as nonsocial ones.

Why should dogs be any different? If anything, the research was showing that dogs care intensely about the meaning of human signals. In light of Hare's findings, it seemed likely that Callie looked at my hand signals and constructed a dog theory of what I was thinking or at least intending. *Dog theory of mind.*

And if Callie was trying to intuit what I was thinking, it was inevitable that I would do the same and try to intuit what she was thinking. Locked in our MRI pas de deux, staring into each other's eyes, I had had the overwhelming sense that we were directly communicating our intentions to each other. Callie's caudate activation was just the first piece of evidence that my intentions had been received, and understood, in her brain.

Dogs, like humans, just want to be understood. Proof of actual mentalizing, though, would take some further examination of the brain activation in regions outside the caudate.

In fact, Callie showed evidence of more than reading our human intentions. She indicated her intentions. At dinner, she stood in front of the glass door leading from the kitchen to the back porch. She turned her head and looked at me. Then she turned back to gaze longingly outside. Back to me. *Come on, I want to go outside.*

She didn't bark. She didn't scratch at the door. Callie clearly communicated her intentions with her eyes. Just like humans.

I let her out, and she went racing through the ivy after some animal.

Callie's behavior may seem unremarkable. She had probably been doing things like this for as long as she lived with us, but I had never had reason to pay much attention to the nuances of what she was doing until now. But with the results from the Dog Project, it now became a matter of scientific interpretation. Either she was a Pavlovian learning machine — great at making associations between events but without interpreting them — or Callie was a sentient being who understood, at some level, what I was thinking and reciprocated by communicating her thoughts within her behavioral repertoire.

I suspected the latter, but the proof was still hidden in the fMRI data.

Callie gave up whatever she was hunting. Long ago, she had quickly learned how to work door latches. Whether it was from luck or from watching humans, I don't know, but now, she ran full tilt and jumped to push open the porch door, precisely timing her leap to hit the handle. She blasted into the kitchen with a burst of energy.

She immediately went over to Helen and rested her head on Helen's thigh.

"Look!" Helen said. "She's doing the 'touch' command."

"She's telling you something," I said.

"She wants food?"

"Yup."

Helen laughed and gave Callie a morsel from her plate. I am not sure who was more satisfied: Helen for understanding Callie's intent, or Callie for making Helen do what she wanted.

"I have good news too," Helen said.

"Really?"

Helen paused for dramatic effect.

"Come on," Kat said. "Don't keep us hanging."

"I got an A on my science test."

"Yay!"

"That's awesome," I said. "I'm very proud of you. You had to work really hard to do that."

Helen beamed.

Sometimes playing hooky really does pay off.

20

Does My Dog Love Me?

THE FIRST PHASES OF THE Dog Project were coming to an end. Callie had been to the scanner four times and McKenzie three. We had proved that the dogs could hold still enough to obtain high-quality images of their brains. And even more impressively, we had shown that the reward systems of their brains activated in response to the appropriate hand signals. We had finished the first scientific paper and sent it off for publication, which meant that we had some time to reflect on what we found and what we wanted to do next. As far as we knew, we had the only two dogs in the world trained to go into an MRI scanner, and we had proven that the whole crazy idea wasn't so crazy after all.

The excitement in the lab was electrifying. I had been scanning human brains since fMRI was discovered, but nothing I had experienced in my career matched this intensity. Even the dawn of human brain imaging didn't match. Perhaps because scientists had been studying the human brain in various ways for over a century, we already knew a lot about how it worked. More often than not, brain imaging tended to confirm what we knew about the human brain and rarely resulted in a sea change in our understanding of the human mind.

But the Dog Project was entirely different.

I felt like Christopher Columbus discovering the New World. The dog's brain was a great, unexplored continent. We had no idea how the canine brain worked, but we had the tools to figure it out and two subjects ready to assist. All we had to do was step into the unknown and start exploring.

The screensaver on Lisa's computer was displaying a montage of Sheriff. Sheriff was almost two years old. Lisa had acquired him as a puppy when she graduated from Emory and started working in the lab. He was the first dog she could truly call her own, and she absolutely adored him.

"You really love Sheriff, don't you?" I commented.

"Of course," Lisa replied, "and he loves me too."

Gavin, who had been observing with bemusement, couldn't resist teasing Lisa about this.

"That depends on what you mean by love."

Lisa, ever the pragmatist, replied, "Love? I would accept codependency." She was dead serious. "Look, I think the best you can hope for with humans is to eventually have a relationship where both people are mutually dependent on each other. What's wrong with that?"

She had caught Gavin uncharacteristically off guard and he had no response. Lisa continued. "So what if Sheriff's love for me is based on food and belly rubs? He gives back affection and companionship. If most human relationships were that simple, more people would probably be happier."

"What if we could prove that Sheriff loved you?" I asked.

"You mean more than food and belly rubs?"

Gavin rolled his eyes and said, "That's impossible."

Andrew, who had refrained from wading into the debate on love, had been staring intently at his computer screen. "Check this out."

On the screen was the structural image of Callie's brain. I had now seen this image a hundred times and knew it better than my own brain. Overlaid was an activation map. We had been looking at pictures like this for weeks and I had become accustomed to seeing the red, orange, and yellow hot spots superimposed on the caudate nucleus — the center of the reward system. But this image was different.

Andrew had digitally warped McKenzie's brain to match Callie's. This is a normal step in the analysis of human fMRI data. When we collect data on a large number of subjects, we need a way to compare activation in everyone's brains. But because every person's brain is physically different, we use a digital method that morphs each brain into the same size and shape. This allows scientists to average the activation patterns of many individuals and determine the commonalities of brain function.

In humans, brain sizes tend to vary by about only 1 or 2 percent. Some people have round heads while others are more oval-shaped. Even so, the basic anatomy is pretty much the same, and we need to stretch and twist the brains only a little bit to make them all match up.

Dogs are different. Of all the species on the planet, dogs have the largest variations in size. What other species can range in size from a 4-pound Chihuahua to a 150-pound Great Dane and still be considered the same animal? As you might expect, their brain sizes have a similarly large variation.

When we started analyzing the data from the Dog Project, we did it separately for Callie and McKenzie. McKenzie was about 50 percent larger than Callie, so we knew their brains were going to be different. Because of this large variation in size, we didn't think the usual computer algorithms would work, so we hadn't even attempted to digitally combine their brains.

Until now.

By carefully identifying key landmarks in the dogs' brains, Andrew had been able to get them to line up. Once aligned, he was able to perform an analysis on the combined dataset. They say that two heads are better than one, and in this case that was absolutely true. Although both Callie and McKenzie had performed beyond our expectations, they still had their limits. They had each stayed in the MRI for ten minutes of continuous scanning. But ultimately, the noise and confinement wore them down, and they got tired of the task. In the end, Callie had sat through almost forty repetitions of the task and McKenzie about thirty. This was good enough to prove the feasibility of canine fMRI. But to go to the next step, and really start figuring out how the dog brain worked, we needed a lot more repetitions and, ideally, a lot more dogs. Combining the results from Callie and McKenzie was a first step in this direction.

More observations meant more power to detect faint signals in the brain. By merging the datasets of the two dogs, we were now staring at a result on the computer screen that we hadn't seen when looking at the dogs individually.

Andrew pointed to an area of activation on the side of the brain. This region was about a centimeter higher than the reward system, and it was located in the middle of the cortex. Since the usual landmarks of the human brain didn't apply, we were left guessing what part of the dog brain we were looking at.

Cross-referencing an atlas of dog brain anatomy, I asked, "Is that the motor cortex?"

Andrew shrugged and said, "It's in the middle of the cortex, about where the human central sulcus would be."

But the dogs weren't moving in our experiment. Why would we see activity in the motor area?

"Mirror neurons," I said.

Mirror neurons are a specific type of neuron in the brain that fires both when an animal initiates a movement and when it observes the same type of movement in another animal. They were originally discovered in the early 1990s by researchers recording the brains of monkeys. The scientists were primarily interested in how the motor system functioned, especially when the monkey decided to reach for an object. They implanted electrodes to record from the area of the brain just in front of the central sulcus, called the *premotor area*. These neurons did, in fact, begin firing just before the monkey moved its hand. Somewhat accidentally, though, the scientists also noticed that these neurons fired when the researchers reached into the cage to replace the object that the monkey was trying to get, even though the monkey wasn't moving at that moment. They were dubbed mirror neurons because they seemed to mirror both observation and action. They fired when the animal initiated a motor act as well as when somebody else performed a similar action, and it didn't seem to matter whether it was a monkey or human hand that was doing the reaching.

It wasn't long before researchers began searching for mirror neurons in humans. Using fMRI, several experiments found evidence for the same mechanism operating in the premotor area of the human brain, as well as a number of other areas. Rather than controlling the movement of a particular part of the body, these mirror neurons seemed to control action *goals*. For example, a baseball pitcher tries to throw the ball in the strike zone. The mirror neurons in a pitcher's brain don't control the muscles of the arm directly. Instead, they act like a guidance system so that all the muscles of the body act together to reach the ultimate goal of depositing the baseball in the catcher's mitt at the desired location. And if a pitcher watched someone else doing the same thing, the pitcher's mirror neurons would fire while he observed — as if his brain were simulating the act of pitching.

The interest in mirror neurons continues to intensify. At a basic scientific level, these neurons seem to play a key role in linking action production with action observation and to allow animals to understand the actions of other members of their species from their own perspective. Many researchers have suggested that mirror neurons are the basis of empathy. If this turns out to be true, then mirror neurons not only allow us to simulate the actions of each other from the inside, but they may allow us to feel what someone else feels too.

The role that mirror neurons play in feeling empathy continues to be debated, but the evidence suggests a route to empathy through imitation. Humans, in particular, have strong innate tendencies to imitate each other. When someone smiles at us, we can't help but smile too. This type of imitation seems to be wired from birth. Infants smile in response to adults smiling at them and also initiate smiles to receive the same response from their parents. The mirror neuron system, by serving as the link between observation and action, may control this type of imitative behavior.

It is through imitation that we begin to feel what someone else feels. Several experiments have shown that the more people imitate each other, the more empathic they become. Although it remains to be proven that mirror neurons are the basis for empathy, it does seem clear that they play an important role in the precursors to empathy. Without the mirror neuron system, it would be unlikely that people would have any empathy at all.

Apart from monkeys watching humans reach for stuff, nobody had demonstrated cross-species mirror neuron activity. Even with the monkeys, a human hand looks an awful lot like a monkey hand. They both have four fingers and an opposable thumb.

But dogs don't have thumbs. They don't even have hands.

And yet Callie's and McKenzie's motor cortices were activating in response to our hand signals. They weren't moving, so maybe this represented mirror neuron activity. But this would be considerably more complex than monkeys observing human hands. If the activity we found came from mirror neurons, this would mean that the dogs were performing some kind of action mapping between a human hand and their forepaws. My mind began to spin with the implications.

Dogs walk on their front legs.

But they also use their front legs to do other things. They dig. They jimmy open doors. They swipe food off the counter. And they hold toys and bones with their front paws. Maybe it wasn't so far-fetched that when Callie and McKenzie were watching our hand signals that their brains were somehow simulating actions with their own paws. It would be a way for their brains to translate human action into equivalent dog action.

That would mean that when dogs watched us run, the neurons that controlled running in their brain would start to fire. It would mean that when we ate, their mouth neurons would be going haywire. I knew the absolute truth of this. How many times had I seen Callie licking her chops as I put a morsel of food in my mouth? It was as if she could almost taste it.

If dogs had mirror neurons that responded to human action, did humans have neurons that responded to dog action? Amazingly, yes. In 2010, an fMRI study reported that when people watched silent movies of a dog barking, the parts of the humans' brains that responded to sounds were activated, even though there was no actual sound. It was like the humans filled in the sound of a dog barking just by observation.

But seeing this kind of mirror neuron activity in Callie and McKenzie meant that the whole dog-human relationship was not just a scam. If dogs had the ability to transform human actions into their own doggie equivalent, then maybe they really did feel what we feel. At least a dog version of it.

The caudate activity was proof that we could detect and interpret activity in the dogs' brains. It showed that Callie and McKenzie understood the hand signals for something they liked — hot dogs. But the motor cortex activity suggested that they were more than Pavlovian learning machines. If, as we suspected, the cortex activity was because of mirror neuron activity, here was the first evidence that the dogs might be performing some kind of mentalizing. They were interpreting hand signals and possibly even mapping our hands onto their paws.

It was tantalizing evidence for dog theory of mind.

That evening, I was sitting on the sofa and Callie was doing her usual patrol of the house and yard. Kat and I had taken to leaving the kitchen screen door ajar, even though it let the mosquitoes in the house. It was easier than getting up and down to let Callie in and out. In the distance, I could hear coyotes howling, which normally sent Callie into a barking frenzy.

But not tonight.

After a few circuits of the yard, she came inside and hopped onto my lap. This was unusual because she was never really a lap dog. Mostly she would curl up with Lyra, apparently preferring the contact of her own kind. But tonight she nestled between my legs and laid her head on my thigh. And I was grateful for the dog-human contact.

I stroked her head gently. I loved the way her black fur slicked down on the flatness of her skull. Her eyes began to narrow as she drifted off to sleep.

Did she feel what I was feeling? She could have chosen anywhere in the house to sleep at that moment, but for whatever reason, she chose my lap. It wasn't for food. It wasn't for warmth — Lyra provided more heat than I could. It had to be that she wanted contact with a human. Me. The same desire I had for contact with a dog. Her.

Callie drifted off to sleep. Pretty soon I could feel her legs twitching as she started to dream. I pondered the possibilities of using fMRI to see what was happening in her brain while she dreamed.

My reverie was snapped by the *thwack-thwack-thwack* of her tail on the sofa. She was still dreaming.

Maybe she was dreaming of taking down one of those coyotes. Or maybe catching a tasty rodent in the yard. Or maybe it was just the dogness of being there, in my lap.

And if that wasn't love, then I would surely accept it as a reasonable facsimile.

21

What's That Smell?

THE RESULTS OF THE HOT DOG experiment made me wonder what the dogs thought of us humans. It seemed like there was more going on than just love of hot dogs. Each trip to the MRI, Callie got more and more excited. By the final scan session, she was making a beeline to the portable steps up to the patient table. She would shimmy into the scanner bore even before we had her chin rest in place. Her look said, *I'm ready, let's go!* She liked interacting with all the people, and, as everyone agreed, she liked showing off. She had become a diva.

Callie had also gotten used to McKenzie. If I had to characterize their relationship, I would call it one of mutual nonthreatening coexistence. We used the lab as our staging area before walking everyone across the campus quad to the MRI facility. Callie and McKenzie would greet each other in the lab with a perfunctory butt sniff and tail wag. That would usually be it, as both dogs preferred checking out the humans in the room. Once at the scanner, though, Callie would start getting more excited. If it was McKenzie's turn to go into the MRI, Callie would climb up on the patient table and try to get into the MRI before McKenzie. Callie would have to be carried off the table and kept in the control room while Melissa and McKenzie got situated.

I found this behavior fascinating. It seemed clear to me that the dogs treated each other differently from the way they treated us humans. This is despite the popular notion that we humans are a "pack" to the dogs — a sort of extended doglike family.

This gave me an idea for another fMRI experiment.

How do dogs categorize humans? Either dogs have separate categories for dogs and humans, or they lump us together as either pack or not pack.

To my eyes, Callie and Lyra behaved like pack mates. They ate together. They slept together. And they played together. It was no different from what we humans in the house did with them. And while we viewed them as family members, it would be nice to know if they viewed us that way too. Because Callie and Lyra were unrelated, and they were obviously unrelated to us humans, the notion of a pack would have to be what anthropologists call *fictive kin*.

Humans are particularly good at treating genetically unrelated friends as if they were family, especially if they go through an intense experience together. This is why soldiers call each other "brother." If people do this, maybe dogs did too. If dogs viewed their humans as part of their pack — a sort of extended family — then dogs and humans should result in similar activation in the dogs' brains.

So what might distinguish humans and dogs — at least in the dog's mind? Apart from appearance, the most obvious is smell. After a dog sees another dog, it will make a visual assessment of body language, like how the tail is held, and decide whether to approach. If it does approach, then they will sniff each other. It is similar when dogs see humans. After a visual assessment, a dog will usually approach and scent the person.

A dog's sense of smell is about one hundred thousand times as sensitive as that of a human. They also have an additional structure, a fluid-filled tube called the *vomeronasal organ* (VNO), thought to be specialized for detecting scents from other dogs and therefore to function in some capacity for social signaling.

With that powerful of a sense, you can be sure that a large portion of the dog's brain is devoted to processing smells. Even so, I was still shocked when we got the first images of Callie's and McKenzie's brains. Where we would normally see a big frontal lobe in humans, the dogs had almost nothing. Instead, extending toward their snout, was a massive phallic protuberance — the olfactory bulb. A rocket in the socket. Humans have nothing like that. And it accounted for about 10 percent of the dog's entire brain.

We usually think of smell as one of the five senses and a generally passive process. Odorants drift into our noses; receptors detect them and send signals to our brains. However, more so than vision or hearing, smell is an active process involving many groups of muscles. Animals can control the rate at which odors enter the nose by the way they sniff. Sniffing involves muscle movements in the face and the nose. It requires movements of the diaphragm to control the rate of air movement. And there is likely some control over the fine hairs inside the nose. This means that for smell in particular, we would also expect to see the involvement of parts of the brain that control movement.

If the scent of a dog activated the brain in the same pattern as the scent of a human, then that would tell us that dogs lumped us in the same category as them. If, on the other hand, dog and human scents caused different patterns of activation, then we would know that dogs have different categories for us and them.

Like the hot dog experiment, the dogs wouldn't have to do anything except hold their heads still, and they were already pros at that. We would hold up cotton swabs in front of the dogs and let the scent drift into their noses. Later, we could analyze the fMRI data to see which parts of their brains reacted to different scents.

This experiment presented certain logistical complications. Where would we get the scents and how would we get them? These questions became the topic of heated debate in the lab, especially as it became apparent that everyone would have to give something for the cause.

"So let me get this straight," Andrew said. "We are going to present scents from dogs and humans to Callie and McKenzie."

"That's right," I said.

"What kind of scents?"

"Well," I said, "we all know what dogs do when they greet each other."

Andrew didn't like where this was going. "You're suggesting a butt wiping?"

"I don't think there is any other way."

Lisa chimed in and offered an alternative: "Dogs sweat from their paws. You could get scents from there."

"But we would also pick up all sorts of smells from where the dogs walked," I said. "Besides, dogs go right to the butt. As far as they're concerned, that is where the good stuff is."

"Are we talking about a butt wiping or something more substantial?" asked Andrew.

It was a good question. A swabbing of the perianal area would probably do the trick, but there was good evidence that urine would be a more powerful signal. Dogs can differentiate their own urine marks from those of other dogs, suggesting that dog urine contains unique pheromones that are the equivalent of doggie fingerprints.

"I think we need urine," I said.

"What about the humans?" Andrew asked.

"If you're doing it for the dogs," Lisa said, "I think you should do it for the humans."

Andrew and I looked at her aghast.

"What?" she said. "The first thing Sheriff does is stick his nose in someone's crotch."

Although Lisa had a point, there were some boundaries that we couldn't cross. Besides, what she was suggesting could be construed as biohazardous waste by the university lawyers.

"How about a good sweat sample from the humans?" I offered. "As long as they don't wear deodorant, we could have people do a work-out and wipe their armpits with a gauze pad."

There was reluctant agreement with this plan, but this immediately raised the issue of who the canine and human "donors" would be.

The pack versus not-pack question became one of familiarity. To Callie's nose, all the scents in the house were familiar to her: me, Kat, Helen, Maddy, Lyra, and even Callie's own scent. This was her dog-human pack. Melissa and I would be at the scanner already, and our scents would pervade the scanner, setting a backdrop against which other smells could be measured. Ideally, we needed scents from other people in our households to serve as the "familiar human." It would have to be Kat's sweat for Callie and Melissa's husband's for McKenzie.

We would also need a comparison for not-pack, or unfamiliar, scents. We would need scents from strange dogs and strange humans. We drew a chart on the lab wall and listed all the humans in the lab, along with their dogs, and cross-referenced them against whether they had met Callie or McKenzie. Andrew's American Eskimo, Mochi, had never been to the lab. She quickly emerged as the leading contender for "strange dog." Plus, she urinated whenever she got excited, and Andrew would have no problem getting a urine sample from her. Since Callie and McKenzie had met all the humans in the lab, we still needed some "strange humans." Considerable discussion ensued about the logistics of getting fresh sweat on scan day, as well as the possibility that the dogs had already been exposed to the smells of spouses, girlfriends, and boyfriends inadvertently as scents carried on lab members or, as the cops say, "on their person."

In the end, I convinced a neighbor to donate her sweat to be the "strange female" as a control for Kat's sweat. Kat's kickboxing coach agreed to donate his sweat to be the "strange male" control for Melissa's husband.

Timing was critical. Everything depended on getting samples as fresh as possible. For the dogs, that meant morning pee, which, we reasoned, would be the most concentrated of the day. For the humans, they needed to get a good sweat going, which could be mopped up from their armpits. Each of the human donors had been instructed to not shower or wear deodorant for the twenty-four hours prior to sample collection. Everyone was provided with sterile gauze pads, gloves, and a specimen bag to place the sample in.

As usual, the scanner was booked for one p.m. We would need all the samples at the lab by noon so that Andrew, who had volunteered for pee-pad duty, could prepare them for the experiment. He would have to cut the pads into strips with sterile scissors and carefully attach each sample to a six-inch-long cotton swab. Each swab would be numerically coded. This way, neither Melissa nor I would know the identity of the samples and inadvertently cue the dogs during the experiment. Only Andrew would know the code. .

That morning, Kat and I took Callie and Lyra for their walk. I trailed the dogs, looking like a crime scene investigator, wearing purple surgical gloves and toting specimen bags. Callie loved to pee on her walks. As soon as she caught the scent of what I presumed was another dog, she would squat and dribble out some urine. She had a peculiar way of doing it, though. Her bottom never quite made contact with the ground. Instead, she sort of hovered and continued walking, giving the appearance of a duck waddling. Callie never left pee spots. She left pee *trails.*

Her urination habits made it easy to collect a pee sample. As Callie tracked a scent, she intensified her sniffing of a location on our neighbor's lawn. I knew she was about to pee and had my pee pad ready. As soon as she squatted, I thrust the pad beneath her lady parts and was rewarded with a warm yellow stain. Callie looked over her shoulder at me. *Hey! What are you doing back there?*

Lyra was more difficult. The deep fur, stained and matted around her butt, appeared less clean than Callie's. Plus, Lyra peed more conventionally for a female dog: back straight, butt in contact with the ground. The best I could do was wipe her bottom right after she peed. It was enough.

Poor Andrew. We had to lock him in a closet while he cut up all the samples. We couldn't have the dogs getting whiffs of all those great smells before the experiment. After an hour of cutting up urine and sweat pads, Andrew emerged.

"Are you all right?" I asked.

He waved me off. "I just need to get some air."

By now, the excitement of parading the dogs across the quad to the hospital had worn off, and only lab members who actually had a job to do on the Dog Project accompanied us. I still got a thrill out of the walk, though.

Like a well-oiled machine, everyone took up their positions at the scanner. Melissa and McKenzie were there, of course, and they relaxed in the control room until it was their turn. Andrew set up a test-tube rack on a plastic worktable at the rear of the magnet. He inserted the cotton swabs, business end down, in each tube.

For each dog, Andrew had prepared seven swabs: the four combinations of strange and familiar humans and dogs, plus an intermediate category of "acquaintances." Callie and McKenzie were acquaintances. They knew each other, but there was no reason to expect that they viewed each other as part of their pack. We would present their scents to each other for this category. In this fashion, we would have a continuum of familiarity from stranger to acquaintance to household member. Using lab members' sweat, we created the corresponding human acquaintance category. Finally, for a baseline, we used the dogs' own urine as a "self" category.

With Callie in the scanner, the shimming and localizer sequence took less than a minute. She knew the routine. For the functional scans, we modified the hot dog experiment. Instead of holding up a hand signal for ten or fifteen seconds, Andrew would hand me a swab, and I would hold it in front of Callie's nose for a few seconds. She would continue holding still to allow enough time for the hemodynamic response to peak, and then I would reward her with a hot dog treat. It was the same as before except we would insert a smell during the middle of the repetition.

To get her used to a cotton swab being shoved in front of her face, Callie and I had practiced this at home. The first few times she backed away, but she soon realized nothing bad was going to happen and just sniffed.

With the functional scans running, she performed flawlessly. Each swab was presented eight times in random order. It took two functional runs lasting six minutes each, and then we were done. Another four hundred scans in the can.

McKenzie, on the other hand, was not having a good day. She did not like the smells or the swabs coming at her. From the control room, I could see that her brain images were moving. Even though we acquired nearly five hundred images, most were unusable. She would have to come back another day after more training with the swabs.

• • •

Like we did in the hot dog experiment, Andrew and I analyzed the smell data individually for the two dogs as well as combining their brains. Both analyses yielded surprising results. When we combined their brains, we were able to identify the parts of the brain that reacted to the different smells in both dogs. This showed the common regions of activation. In contrast, the individual analyses told us how the dogs reacted differently. We focused on two comparisons.

First, we compared the brain activity of dog scents to human scents. This was done by ignoring whether the scent was familiar or strange. We simply averaged all the dog scents together and all the human scents together and compared the two brain patterns. The first thing that popped out was that the canine smells strongly activated the olfactory bulb and the frontal cortex above it. I suspected that this was because dog urine is a more potent stimulus than human sweat.

When we compared the familiar scents to the strange scents, ignoring whether it was from a dog or a human, once again, we found more activation of the olfactory regions to the strange smells. This demonstrated that the olfactory activation is controlled not just by the potency of the smell but also by its familiarity. Familiar smells don't require much brain processing. Strange ones do. Consistent with this interpretation, the dogs' own urine didn't evoke any detectable brain activity. Just like humans aren't aware of the smell of their own breath, dogs seem to tune out the smell of their own pee.

Strangely, we also observed strong activation to the unfamiliar smells in the cerebellum, a part of the brain usually associated with movement. When I was presenting the cotton swabs to Callie, sometimes she sniffed more intensely. The cerebellum activation was most likely the neural origin of the sniffing, which would be more intense for smells the dogs hadn't encountered.

The most interesting finding appeared when we subdivided the dog and human scents into their subcategories of familiar and unfamiliar. One, and only one, type of smell activated the caudate: *familiar human*. This was especially true for Callie. In her case, the familiar human was Kat.

Kat's sweat activated Callie's caudate — same as the signal for hot dogs. But Kat wasn't even at the scanner. This meant that Callie had identified the scent as Kat even though she wasn't physically present. And if Callie had a mental category for Kat that didn't require her physical presence, then this suggested that Callie had a sense of permanence for the people in her household. She knew who her family was, and she remembered them. We found further evidence for this interpretation in an area called the inferior temporal lobe. This part of the brain is closely associated with memory function, and like the caudate, the inferior temporal lobe was strongly activated by the smell of a familiar human.

The inferior temporal activation told us that the dogs remembered their human family, and the caudate activation, more prominent in Callie, told us that her remembrance of Kat was a positive one. Could it be longing? Or love? It seemed entirely possible. These patterns of brain activation looked strikingly similar to those observed when humans are shown pictures of people they love.

The results of the smell experiment expanded our understanding of the dogs' mental world. All through the Dog Project, we had been focused on the nature of the dog-human relationship. We love dogs, but what do they think of us? Even with just two dogs, a picture was beginning to emerge. The pattern of activations in the cortex suggested that they concocted mental models of our behavior, which might be due to mirror neuron activity. But regardless of the mechanism, the smell data showed that their mental models included the identity of important people in their lives that persists even when the people aren't physically present.

I was willing to accept that as an acceptable demonstration of love from Callie. But even if I was being too generous, the fact that the dogs knew who we were, and that they had categories for us, indicated that we humans make a lasting impression on our dogs. We are appreciated.

22

First Friend

WHEN WE BEGAN THE DOG PROJECT, we had no idea what we would find. What started as a half-baked idea to scan dogs' brains mushroomed into a full-fledged research program faster than I ever expected. Even with just hot dogs, and then smells, we had found evidence that dogs mentalized about the humans in their lives. I suppose this should not have been surprising. Many dog owners are convinced that their dogs know who they are and return their love for them. But, for the first time, we saw direct evidence of reciprocation in the dog-human relationship and social cognition in the canine brain.

This was truly exciting, but, in the interest of scientific objectivity, we had to be careful in generalizing from our experiments. The World Health Organization estimates that the population of dogs is 10 percent that of humans. That translates into roughly 700 million dogs worldwide. And we had studied the brains of precisely two of them. Although we had expanded the enrollment in the Dog Project since our initial experiments, we were still studying a very selective group of dogs. These were dogs that were loved by their humans. But even that is not enough. Most dogs aren't willing to go into an MRI, and most people aren't willing to train them to do so. That still leaves the 700 million dogs of the world. What did our experiments tell us about those dogs and their relationships to humans?

From an evolutionary perspective, dogs are incredibly successful. Their numbers speak to that. Given that dogs share their environmental niche with humans, their success must be a result of learning how to read us. Not just reading human behavior but, I believe, learning to read our intentions, which means that they have a theory of mind for humans. And that is exactly what we found in the Dog Project. So even though Callie and McKenzie were rarified representatives of the world dog population, what we found in their brains showed the defining characteristic of dogs: social learning. Their brains showed that they cared about human intentions.

Proof of social cognition means that dogs aren't just Pavlovian learning machines. It means that dogs are sentient beings, and this has startling consequences for the dog-human relationship.

Most of the dogs in the world are village dogs. They are not anyone's pets, although they might look that way at first glance because they often gather near humans. People know who the dogs are, but they might not have names. Village dogs insinuate themselves into the fabric of human societies. They feed on scraps, sometimes garbage, and sometimes food that is deliberately left by humans for them to eat. In some parts of the world, people let them hang around just so that they can eat them later.

If Callie had lived anywhere else in the world, she would have been a village dog. She had that rangy appearance — not too big, not too small — and the eyes of an opportunist. For the first year that she lived in our house, I was convinced that she would run away at the first opportunity of something better. But after the Dog Project, I no longer thought that. The project had changed both her brain and mine.

Indeed, if there is one thing dog ethologists can agree on (and it might indeed be only one thing), dogs are masters of change. If nothing else, dogs' defining characteristic is their adaptability. Apart from vermin, dogs are the only mammalian species that is found everywhere humans are, and humans have occupied every habitable niche on the planet. As the ethologists Raymond and Lorna Coppinger observed, "The rapidity with which the dog has changed form, and the seemingly endless varieties of its form, challenge the theory of Darwinian evolution, that adaptation must be a slow process." The Coppingers were referring primarily to dogs' changing physical form, but the same can be said for dogs' behavior.

When scientists speak of behavioral change, they are really talking about learning. And, as far as we know, there are only two mechanisms of learning that animals employ: associative learning and social learning. For a century, Pavlovian behaviorists have argued for the predominance of associative learning. Animals, dogs included, are great at learning associations between neutral events and things that they like, such as food, or things that they don't like, such as pain. But associative learning cannot explain all of animal behavior. For one, it is inefficient. For an animal to learn associations, it has to actually experience the events. This is a trial-and-error process. By this learning mechanism, a dog would actually have to touch its paw to a hot stove to learn that that is something to be avoided.

Social learning is far more efficient. Many animal species employ social learning. Songbirds, for example, learn their species-specific calls from each other. But besides humans, dogs may be the best of all. By watching other dogs, Fido can learn a great deal. He doesn't have to burn his paw to learn that the stove is dangerous if he sees another dog (or human) do the same. And, of course, puppies learn from each other and their mother, copying behaviors like pulling toys.

I have often wondered how dogs got so good at social learning. While many animals learn from members of their own species, dogs are one of the few that can learn from *other* species. Herding dogs, for example, learn by observing sheep and cattle. And all dogs learn by observing humans and other members of their households, just like Callie learned how to open doors. Village dogs, even though they are not attached to specific humans, exemplify this ability for social learning. There is no other way they could keep up with the ever-changing form of human society.

In the hot dog experiment, we found that the meaning of the hand signals had transferred to the caudate — a brain region known to be associated with positive expectations. While a cool scientific finding, it was not really unexpected given what we had known about Pavlovian learning. What was more revealing, and what we never commented on in our academic papers, was all the other stuff going on in Callie's and McKenzie's brains. The motor cortex activity. The inferior temporal lobe. Those were the regions that pointed toward a theory of mind, and they were the same regions that popped up in the smell experiment associated with familiar humans.

These cortical regions showed that the dogs might be constructing mental models of our actions. The inferior temporal lobe suggested that they were recalling memories, perhaps what one hand pointing up meant, or the identity of the person associated with a sweat sample. These are the types of mental processes that any sentient being would use on a daily basis. Humans use memories and ascribe meaning to people and actions all the time. Apparently, so do dogs.

Even though we found evidence for canine theory of mind in our experiments, Callie and McKenzie were not exactly the same in this regard. They showed differences in how their brains reacted to the hand signals and to the smells. With only two subjects, it is difficult to draw sweeping conclusions, but I will take scientific license to speculate.

In the hot dog experiment, McKenzie had stronger caudate activation to the "hot dog" hand signal. Strangely, Callie was the food lover, while McKenzie much preferred toys as rewards during training. Because of her great love of hot dogs, I had expected Callie to show the stronger caudate activation. But she didn't. One possibility is that because Melissa and McKenzie had competed in agility competitions, McKenzie was more attuned to hand signals. I had not taught Callie a hand signal before the Dog Project, which might have put her at a relative disadvantage. Another possibility, which I think is very likely, is a genetic basis.

Even though we thought she was a feist, Callie was more like an adopted village dog. A mutt. McKenzie, on the other hand, was meticulously bred to be a herding dog. Border collies are known for their stares, what the Coppingers refer to as the *eye-stalk*. Border collies don't just see with their eyes; they use them to control other species. There may have been much more going on in McKenzie's brain as she not only interpreted Melissa's signals but also returned them with her eyes. While I had noticed a flicker of that in Callie when her eyes dilated in anticipation, it was nothing like being stared down by a border collie.

In the smell experiment, though, the pattern reversed. Callie had the stronger response to the familiar human smell. Maybe that was because Callie slept with Kat and me in our bed, while McKenzie slept in her crate. Or maybe the bond between Callie and Kat was stronger than that between McKenzie and Melissa's husband. Could it be that our dogs tell us more about our human relationships than we tell ourselves? The term *therapy dog* would take on a new meaning.

The evidence for social cognition in dogs' brains has important implications for the dog-human relationship. Dogs watch us constantly, even though we may not be aware of it. With furtive glances, they take in their surroundings and form mental models of what we humans are intending to do. It is the humans who are unaware of the dogs. And that is where misunderstandings can arise.

Humans are sloppy creatures. Like the proverbial bull in a china shop, we are oblivious to our body language. We bump into objects. We accidentally step on our dogs' tails. We emit a constant stream of sounds with frequently inconsistent meanings. It is a wonder that dogs can pull anything consistent out of this barrage of signals. And yet they do.

The whole purpose of the Dog Project was to understand the dog-human relationship from the dogs' perspective, and the most important thing that we learned was that dogs' brains show evidence of a theory of mind for humans. This means that they not only pay attention to what we do but to what we think, and they change their behavior based on what they think we're thinking. They are the Zeligs of the animal kingdom.

Zelig was a fictional character created and played by Woody Allen in his 1983 movie of the same name. Zelig had no personality of his own. Instead, he took on the personality and physical form of people around him. Because doctors thought he was crazy, Zelig was institutionalized in a mental hospital, where he took on the form of a psychiatrist. (His real psychiatrist, a woman played by Mia Farrow, falls in love with Zelig and they eventually run off together at the end of the movie.) Apart from being a terrific film, *Zelig* is a case study in theory of mind. Zelig's problem was that he had no sense of self. He had a sense only of others. The sense was so strong that he knew what was in other people's minds, and he became them.

If dogs are like Zelig, then the form they take depends on the people they live with. If they live with calm, consistent humans, they will pick up on those qualities. If they live with people who talk constantly, without saying anything, dogs will quickly learn that there is no useful information in their chatter. With their social cognition skills, dogs do not need an excess of jabbering. Patricia McConnell, the well-known animal behaviorist, has written extensively about the effectiveness of the less-is-more approach to dog communication. The takeaway is that humans should pay more attention to what their body language communicates than what their mouths say.

Dogs' sensitivity to social signals also puts a new twist on the old notion of human as "pack leader." While it is easy to confuse being a pack leader with being dominant, that is a mistake that has harmed more dogs than any other piece of advice.

The better analogy for being a pack leader comes from management literature. While there are different styles of leadership, the most important characteristics of a great leader are clarity and consistency. Without those two qualities, people (and dogs) cannot know your intentions. Great leaders are also respected, not because of their position, but because of their inner strength and integrity. Leaders do what they say. Leaders listen to people, and although they may not always agree, they have respect for others. Great leaders help people.

It wasn't until I started working with Callie in the Dog Project that I realized just how much she could be dialed in to my signals. Like a catcher and pitcher, we became a team. She had always had that ability. I just hadn't given her any clear direction before.

Eventually, I came to the conclusion that the key to improving dog-human relationships is through social cognition, not behaviorism. Positive reinforcement is a shortcut to train dogs, but it is not necessarily the best way to form a relationship with them. To truly live with dogs, humans need to become "great leaders." Not dictators who rule by doling out treats and by threatening punishment, but leaders who respect and value their dogs as sentient beings.

Even though I couldn't have known about the depth of dogs' social cognition when we started, respect for dogs had been built into the Dog Project. Early on, we had made the presumptive decision to give the dogs the right of self-determination. If they didn't want to be in the MRI, they could walk out. Same as a human. We created a consent form. Although the dogs did not have the capacity to understand its contents, their human guardians did. The guardians were able to weigh the risks, however minimal, against the benefits and decide whether it was in the dogs' best interest to participate. The legal model we used for this process was lifted from the manual on human experimentation. We treated the dogs as if they were human children. But nobody had ever done this before. In the eyes of the law, dogs are still considered property.

The brain-imaging results showed that dogs had mental processes substantially similar to our own. And if that is true, shouldn't they be afforded rights similar to humans? I suspect that society is many years away from considering that proposition. However, recent rulings by the US Supreme Court have included neuroscientific findings that open the door to such a possibility. In 2010, the court ruled that juvenile offenders cannot be sentenced to life imprisonment without the possibility of parole. As part of the ruling, the court cited brain-imaging evidence that the human brain is not mature at age thirteen, supporting the notion that children, even teenagers, are not fully responsible for their actions. Although this case has nothing to do with dog sentience, the court opened the door for neuroscience in the courtroom. Perhaps someday we may see a case arguing for a dog's rights based on brain-imaging findings.

Many people will find the argument for dogs' rights troubling. After all, most dogs of the world are cared for by no one. Perhaps a fifth of the world dog population is lucky enough to live in the company of humans, and some fraction of those dogs actually live a comfortable life. Most people just don't care about dogs.

But if dogs have more capacity for social cognition than we previously thought, then we must reevaluate where they belong on the spectrum of animal consciousness. And this necessitates a reevaluation of their rights. Dolphins, whales, chimpanzees, and elephants, for example, have all been recognized as having substantial cognitive capacities, even self-awareness, and as a result are increasingly being protected from hunting (although many people do not recognize these protections). Throughout human history, there has been an undeniable trend toward granting basic rights of self-determination and liberty to groups of people that were once thought inferior. People of color, women, and gays and lesbians have all benefited from a general recognition of equality.

Will animals be next? Because animals cannot speak, it will take a technological revolution like brain imaging to show that they have many of the same mental processes humans do.

Unfortunately, scientists will continue to resist the obvious. Many scientists rely on animals for experimentation. The animals, of course, have no choice in the matter. It is terminal for them. Even within the small group of scientists who have since begun using MRI to study dogs' brains, there is still a general disregard for the dogs' welfare. By buying "purpose-bred" dogs, many of these labs continue to support the disgusting industry of breeding dogs specifically for research. And, to my knowledge, my lab is still the only group that cares enough about its canine volunteers to go through the considerable effort of training them to wear ear protection.

We still need animals for research. But the vast majority of this research is currently for humans' benefit. We need less of that and more research that directly benefits the animals themselves. Let's start with dogs.

23

Lyra

THE SCIENTIFIC PAPER DESCRIBING our first results with the hand signals was published on a Friday afternoon. The event signaled the conclusion of the first chapter of the Dog Project. For the first time in months, I had a weekend with nothing to do, and I planned to take full advantage of the leisure time.

It was May in Atlanta — one of two perfect times of the year, the other being in October. In these months, and only in these months, the atmosphere achieved a momentary stability as the air from the Gulf of Mexico was perfectly balanced by fronts settling in from the north. The air was warm but not humid. The pollen had disappeared. And the city was lush with new growth.

I lounged on the porch and enjoyed the spring air while Callie bounded in and out of the house with her favorite toy — a blue Kong. The Kong, shaped like a snowman, was just the right size so that Callie could get her mouth around the small end. Amazingly, the squeaker was still intact. She loved to carry it around, teasing me to take it away from her and darting away as soon as I got close. As I dozed off, I could hear her in the distance working the squeaker. The hours slipped away.

Helen woke me.

"Dad," she said, "Callie is whimpering."

Callie was in the family room, still chewing and squeaking her Kong. She appeared fine. Except she was chewing and making little whining noises.

I wrestled away the toy and threw it in the other room. Callie retrieved it and settled down just out of my reach, per her usual game. She continued to chew and whine. Callie was not generally a whiner. Apart from the time when she ate her way into the emergency room, I had never heard her complain about anything. Strangely, she seemed fine. I shrugged and told Helen not to worry.

"Maybe she's inventing a new game."

I returned to the porch to resume my nap, and Helen went back to playing a video game.

Soon, the sun dipped behind the tall southern pines, signaling the dogs' feeding time. Callie had stopped chewing and whining and was asleep on the couch. Normally, Lyra would be right there in the kitchen barking up a storm to feed her. I called for her but got no response.

It didn't take long to find her. She was in the living room, panting heavily. A pile of foul-smelling poop lay next to her.

Oh, Lyra, I thought, *another accident.* For the last several months, Lyra had intermittently had some mild digestive issues. Maybe once a week she would urp up a small amount of yellow stomach fluid. It never seemed to bother her, and she would always eat normally. It is a fact of living with dogs that from time to time you share your home with their stomach contents. Newton used to love chewing off the tags from articles of clothing. This inevitably resulted in him vomiting a few hours later. You get used to it.

While Kat cleaned up the mess, I fed Callie.

I had assumed that Lyra would hear the food in the kitchen and appear shortly. When she didn't, I went to check on her.

She was lying on her side. I rushed to her and stroked her head. I didn't want to upset the girls. Lyra's eyes were open, but she wasn't focusing on anything. Her breathing was rapid and shallow. I buried my face in her ear, whispering her name and trying desperately to suppress my rising sense of panic. But as soon as I did, I could feel that her lips and nose were cold. Her gums were pale. I ran to get Kat.

"Something is seriously wrong with Lyra," I told her. "We have to get her to the emergency room right now."

While Kat got a towel to lift Lyra, I broke the news to Helen.

"Helen, Lyra is really sick." Fighting back tears, I went on. "We have to take her to the vet right now." Helen immediately sensed the seriousness of the situation.

"Can I come?" she said.

"Yes, of course."

"Is she going to be okay?"

Tears started down my cheek. I hugged her.

"I don't know."

Kat and I rolled Lyra onto a beach towel and carried her to the minivan, where we carefully placed her in the back. Helen sat down next to her and stroked her head. It was all a bit much for Maddy, who asked to stay at home. Kat agreed to stay with Maddy while Helen and I sped off to the ER, only five minutes away.

It was an early Saturday evening and a crowd of people, cats, and dogs was filling the vet ER. An old man was trying to sign in a schnauzer. I ignored him and demanded immediate help.

"How much does your dog weigh?" asked the receptionist.

"Eighty pounds."

"Two techs to the front desk for immediate assistance!" she barked into the PA system.

In less than a minute, two women appeared with a gurney, and we rushed to the parking lot. I opened the tailgate of the minivan. Helen was still sitting with Lyra. I could tell from the techs' facial expressions that this was not good.

"How long has she been breathing that way?" one asked.

"Less than an hour," I said.

They lifted Lyra onto the gurney and rushed her into the back of the hospital. Helen and I sat down in the waiting room. Numbly, I pulled her tight.

We didn't have to wait long. Another young woman, with long blond hair and kind eyes, introduced herself.

"I'm Dr. Martin, the staff veterinarian tonight."

I looked at her, fearing the worst.

"Lyra's blood pressure is extremely low, and we can't get an IV started in any of her paws," she explained. "We need to make a cut in her neck and put the IV there to give her fluids. Is that okay?"

I said yes, and she rushed away.

The receptionist motioned me to the front desk to sign paperwork. Having been there before, I knew they wanted me to guarantee payment. Of course I would. The last form, though, I was not prepared for. Did we want CPR performed if Lyra's heart stopped? If not, then she would be a DNR: do not resuscitate.

Even in humans, CPR offers a fifty-fifty chance at best. If Lyra's heart stopped, that could mean chest compressions, defibrillation, intubation, even open cardiac massage.

I called Kat.

"They want to know if she should be DNR," I said.

"What's wrong with her?"

"She's in shock, but they don't know why," I said. "They're doing a neck cut-down to get fluids into her, but they need to know if we want them to do CPR if her heart stops."

Kat was an ICU nurse. She knew what was down that road.

"I don't want her intubated," she said. "I don't want her to suffer."

I didn't either. I checked the box for DNR and sat down with Helen. After fifteen minutes the vet came out and explained the situation. They had managed to get an IV into Lyra's neck and she seemed to be responding to the fluids they were giving her. Her blood pressure, though, remained unstable. The lab work showed that the level of potassium in her blood was elevated. Everything else was normal.

"Does she have Addison's disease?" the vet asked.

Addison's disease, technically called adrenal insufficiency, is a somewhat rare disease in both humans and dogs in which the adrenal glands cease to function. The adrenal glands sit atop each kidney and produce several hormones that are necessary to maintain vital functions of the body. Adrenaline is produced there and helps maintain blood pressure and heart rate. The adrenal glands also produce hormones that allow the body to absorb sodium from food. Nobody knows what causes Addison's disease. It often progresses so slowly, with only the vaguest of symptoms, that it is sometimes never diagnosed. Until the patient enters an Addisonian crisis. A crisis can be triggered by the slightest of stress — a viral illness or even a mild injury. Without the necessary hormones to rev up the body to fight the stress, the patient collapses into shock.

Nobody had ever suggested that Lyra might have Addison's disease. It hadn't occurred to me, Kat, or her regular vet. But the question made me wonder. Lyra had never been a high-energy dog. Could the "Sloth," as we called her, simply have been fatigued and weak? Those would be classic symptoms. The intermittent vomiting could have been a sign too. I didn't know.

Kat arrived and we all went back to see Lyra in the ICU.

She appeared to be sleeping. I was grateful that she didn't appear to be in any pain. Several bags of different fluids were hung on an IV pole. Helen lay down beside her and stroked her head with the tenderest of touches. The vets were giving her steroids, presumptively assuming that she had Addison's disease, but it was all guesswork. There wasn't anything more we could do by staying at the hospital. Lyra appeared stabilized, and our presence could potentially excite her, which could send her into shock again.

I hugged her gently and whispered in her ear, "I love you, Lyra," and wiped my tears on her fur. The vet promised she would call if anything changed.

The five-minute drive back to the house felt like it lasted an hour. None of us said anything.

The phone was ringing when we walked in the door. It was the vet. Right after we had left, Lyra vomited blood and started hemorrhaging from the other end too. If we didn't do something right away, she was going to bleed to death into her digestive tract.

"She has DIC," said the vet. I repeated that to Kat.

Disseminated intravascular coagulation, or DIC, occurs for unknown reasons following trauma or shock. The body goes haywire, clotting in places it shouldn't be and using up the clotting factors in the process. The end result is uncontrolled bleeding, which is what was happening to Lyra. When it happens in people, only the most aggressive care can save the patient, and even then, the prognosis is poor. In the world of veterinary care, DIC is grimly referred to as "dead in cage."

Kat started to cry.

The vet wanted to give her a transfusion of dog plasma, which would contain clotting factors to stop the bleeding.

"Do you think it will work?" I asked the vet.

"I don't know," she said. "Lyra's condition is grave. If we can stop the bleeding, she has a chance." I gave the okay.

"If anything changes, please call us right away."

Nobody wanted to sleep that night. To distract ourselves, we stayed up and watched TV until midnight. Maddy wanted to be alone, and Helen slept with Kat and me. Callie curled up at the end of the bed, confused.

In the morning, I waited as long as I could before calling the hospital. The doctor on call for the day reported that Lyra's lab values seemed stable. Her blood count had not dropped much, indicating that she hadn't lost too much blood. But her clotting factors were still out of whack, and she was still bleeding out of her GI tract. The plan for the day was to try to keep her blood pressure stable.

Around noon the entire family piled into the minivan, and we drove to the hospital. Even Maddy, who normally shied away from intense emotion, seemed to realize that this might be the last time she was going to see Lyra and agreed to come. Her face was twisted up as she tried to contain her feelings.

At the hospital, Lyra was in the same pen as the previous night. She was still sleeping and appeared comfortable. Helen curled up with her, and Lyra sensed her presence. She raised her head and sniffed Helen. The corner of Lyra's mouth turned up ever so slightly in a smile of recognition, and she went back to sleep. Helen covered her up with a blanket that the two of them slept with.

We each took our turns. Watching the girls hug her, knowing in the back of my mind that this could be the last time with Lyra, was the most awful pain. I grieved for Lyra, and I grieved for the girls.

After thirty minutes, Lyra seemed to perk up a bit. She stood up and looked around. Helen's face brightened. But then Lyra shifted position, revealing a bright red stain where her butt had been.

Helen rushed to me, sobbing. I started to cry too.

The vet tech cleaned her up quickly. But since our presence wasn't helping Lyra, we all agreed that it was time to leave.

We tried to have a semblance of normal life at home. Callie seemed out of sorts, wandering the house looking for her big, fluffy pillow. I took to walking her around the block. Usually we walked in the morning and evening, but neither of us could get enough walking while Lyra was in the hospital. By the afternoon, we had been around the neighborhood four times.

I waited until the evening shift at the hospital to call again. Dr. Martin was coming back on duty, and I wanted her opinion of Lyra's condition over the last twenty-four hours.

"She's having runs of v-tach," she said.

Ventricular tachycardia, or v-tach, was a heart arrhythmia. Her heart was racing out of control.

"We just gave her an injection of lidocaine," Dr. Martin explained. "It stopped the v-tach for now."

There was no denying it. Lyra was slipping away. Her heart was racing because her blood pressure was dropping. But when the heart beats that fast there is no time for it to fill with blood, and blood pressure will continue to drop. Maybe she would go on like this for another day or so, but we had to confront the reality that her body was shutting down. Trying to save her would mean multiple drugs, transfusions, and being hooked up to a ventilator. Both Kat and I had seen this happen with people in the ICU, holding off the inevitable while the family held on to unreasonable expectations of recovery.

It was time.

I told Kat what the vet had said. Then we called the girls to the kitchen table and explained Lyra's condition.

"Girls," I began, stifling tears, "Lyra is not doing well, and her heart is struggling to keep beating. It would be wrong to let her go on suffering, just for us."

There was nothing more to say.

It is a heavy burden for an eleven- and twelve-year-old to make a choice between having their beloved dog come home or setting her free from her suffering. To spare them that guilt, Kat and I made the decision for them and simply framed it as the right thing to do. Even if I wasn't sure myself.

I called Dr. Martin and told her that we didn't want Lyra to continue treatment when the prognosis was so poor. She understood and assured me that we were making the right choice.

At the hospital, Lyra looked the same. I was relieved that she still appeared to be sleeping, even though mentally she was probably out of it, bordering on being comatose. Her heart monitor told the story. She was in v-tach, and her heart was beating two hundred times a minute, too fast to maintain blood pressure.

While Kat signed the forms, Dr. Martin explained what would happen next. Helen absorbed the information without expression. We all sat on the floor around Lyra, each of us laying a hand on her. The first injection was an anesthetic. There was no discernible change, confirming that Lyra was already, in effect, asleep, and this knowledge lessened my guilt a little bit. The second injection, a cocktail of chemicals, was just as unremarkable. No shuddering, no movement. Just a cessation of Lyra's shallow breathing. The slight upturn of her mouth — her doggy smile — remained permanently in place.

For the last time, I whispered in her ear so that only she could hear: "Lyra, I'm sorry I let you down. I'm sorry I was deaf to what you were saying. And I'm sorry I didn't understand what Callie was trying to tell me. If only I had taught you to go into the scanner too, maybe I would have known there was something wrong. I will miss you, always."

• • •

By the time we got home, it was dark and it had started to rain. There was no question that Lyra would receive a proper funeral. But I would have preferred to wait until the morning.

Helen summed up the situation: "Dad, I can't sleep knowing that her body is just lying here."

So, with headlamps in place, Kat and I set to the task of digging Lyra's grave in the dark. Despite the rain, the red Georgian clay did not yield easily to our shovels. Neither of us cared. After two hours of digging and prying rocks, we were staring at a hole so deep that we had to stand in it to dig any farther. We both took some comfort in the blisters that had formed on our hands. A tearing of the skin that symbolized the tearing in our hearts.

We lowered Lyra into the hole and called the girls outside.

They each placed a stuffed animal next to her, and we covered her in a favorite blanket. In turn, each of us placed a shovelful of earth in the grave.

The grief was too overwhelming for anyone to speak, so I spoke for all of us.

"Lyra, you were the gentlest, kindest dog we have ever known. You will be in our hearts forever."

Choking back tears, and as I had upon Newton's death two years before, I recited "The Rainbow Bridge":

Just this side of heaven is a place called Rainbow Bridge . . .

24

What Dogs Are Really Thinking

IT HAD BEEN TWO years since the inception of the Dog Project, and our shrine to the dead was now one soul larger. I thought back to the weeks following Lyra's death. Nobody in the house had been the same. Maddy missed cuddling with the big teddy bear, and Kat longed for Lyra's happy, vacant face staring up at her from the foot of the kitchen table. Even Callie had lost a little bit of her spark and had taken to following me around the house. Helen was morose and cried herself to sleep with Lyra's collar in her grip.

After all we had accomplished, I wondered whether Lyra had been trying to tell me something. I supposed it had been possible, but I also knew that her personality was such that even if something had been bothering her, she wouldn't have given any indication. It was the way of the golden retriever. Unflappable and perennially friendly, these are the reasons why goldens are so popular.

But the traits that make goldens so lovable also make it harder to know what they are thinking. I had learned to read Callie but I had taken Lyra for granted. For some time after Lyra's death I faulted myself for this oversight. But gazing at Lyra's picture, I realized just how different our dogs had been. Callie was a hunter. Lyra wasn't. Although Lyra had come from a line of dogs bred for hunting and retrieving, she had never displayed any of those traits. She had never even taken to swimming.

Finally, after two years, the Dog Project had begun to find clues to why we love dogs so much and how dogs became who they are. Eventually, our results might even explain why dogs and humans came together thousands of years ago. The brain data pointed to dogs' unique interspecies social intelligence. In answer to the question "What are dogs thinking?" the grand conclusion was this: *they're thinking about what we're thinking*. The dog-human relationship was not one-sided. With their high degree of social and emotional intelligence, dogs reciprocated our feelings toward them. They truly are First Friend.

Throughout the world, the two most popular pets are dogs and cats, and both are descended from predatory species. It seems odd that the first animals that humans supposedly domesticated were hunting animals. You would think that it would have been much easier for prehistoric humans to take in more docile species. A common explanation for this is that dogs helped humans hunt while cats caught vermin. While plausible, this theory assumes that humans domesticated animals because of their usefulness in survival.

The results from the Dog Project suggest a different explanation. While the caudate activation in the dogs' brains shows that they transfer the meaning of a hand signal to something rewarding like hot dogs, the other brain regions activating point toward a theory of mind. Our results support a theory of self-domestication based on dogs' superior social cognition and their ability to reciprocate in human relationships. Moreover, these interspecies social skills evolved from dogs' predatory past.

Apart from humans, strong evidence for theory of mind has been found in only monkeys and apes, which have social cognition for primates but not necessarily other animals. Dogs are much better than apes at interspecies social cognition. Dogs easily bond with humans, cats, livestock, and pretty much any animal. Monkeys, chimpanzees, and apes will not do this without a lot of training from a young age. And even then, I would never trust an ape.

The different types of social cognition may be a result of the different diets of the species. Apes eat fruits, grasses, seeds, and sometimes meat. Like humans, they are omnivorous. Dogs (and cats), on the other hand, are mostly carnivorous. This means that dogs' ancestors, the wolves, had to hunt their prey. Apart from humans, primates do not depend on meat for a substantial part of their diet.

Hunting is hard. It is not as simple as waiting for prey to wander by. Predator species must outsmart their prey. To some extent, this means that predators must get in the mind of their prey. A lion, for example, stalks a gazelle by anticipating what it is going to do, but the gazelle only reacts. All predators, whether they hunt alone or in packs, had to evolve an interspecies theory of mind to be successful. The brain-imaging results suggested that through evolution, dogs somehow adapted their ancestors' skills in reading the mind of other animals from a predatory capacity to one of coexistence.

Around twenty-seven thousand years ago, a subspecies of wolves domesticated themselves and became dogs. During this period, the ice sheets had reached their greatest extent, stretching as far south as Germany in Europe and New York City in North America. The ice sheets would have pushed humans who had previously migrated north to move south again. The wolves, who were well adapted to cold climates, also would have moved south following the ice sheets. As a result, both humans and wolves probably came into contact with each other more frequently.

Why wouldn't they have eaten each other? Perhaps they did. But more likely, a few wolves realized that they could hang around humans. Some researchers have suggested that the wolves survived by scavenging from human leftovers. However, John Bradshaw has pointed out that wolves require a prodigious amount of food, and it is unlikely that a wolf could have survived exclusively off human garbage. Others have suggested that wolves helped humans hunt. This might have been possible, but even modern dogs need to be trained to help the hunter. And wolves are not nearly as trainable as dogs. Moreover, dogs appear almost nowhere in prehistoric cave art that otherwise depicts human hunting activity.

The results from the Dog Project, however, support a much simpler theory. Because wolves were predators, they were already well evolved for intuiting the behavior of other animals, which meant that wolves had a high level of interspecies social cognition, perhaps even a theory of mind. For wolves used to hunting, it would have been a trivial mental feat to learn the habits of humans. If humans fed them, it would have been simply because they liked having them around, not because wolves provided any survival function. Anthropologists have long known about the universal human tendency to take in animals as pets. Everything from reptiles to birds to mammals. In almost all cases, pets provide no useful function other than it makes humans feel good.

It is not hard to imagine a nomadic tribe of Ice Age humans running into a pack of wolves. A friendlier and more curious wolf might approach the tribe, tentatively at first. A friendly and curious human might leave some food on the perimeter. It wouldn't take long for the two individuals to get close enough to achieve physical contact. Initially, the wolf would probably split its time between the pack and the humans. However, when either the humans or the wolves moved on, the wolf would have to make a choice of whom to follow. It is easy to imagine an exceptionally social wolf, probably a juvenile male, choosing the humans. The human, also probably a child, would see that the wolf was following her and continue to divert food to the wolf.

This scenario, however, would not result in any physical changes in the wolf, at least not for a long time. It is unlikely that an individual group of humans could have supported more than one wolf. As a result, there would have been no opportunity for the wolf to breed with other like-minded wolves. I suspect these "one-off" domestication events happened sporadically throughout the period from twenty-seven thousand years ago until about fifteen thousand years ago. Only when humans stopped being nomadic and stayed in one place long enough to span the reproductive cycle of the wolf did physical evolution start to take off, and the wolf morphed into the dog. The remaining wolves — those who wanted nothing to do with humans — gave rise to the wolves we know today. Modern wolves must represent the opposite end of the canid spectrum from dogs.

The defining trait of dogs, therefore, is their interspecies social intelligence, an ability to intuit what humans and other animals are thinking. Wolves do this to hunt prey. But dogs evolved their social intelligence into living *with* other species instead of eating them. Dogs' great social intelligence means that they probably also have a high capacity for empathy. More than intuiting what we think, dogs may also feel what we feel. Dogs have emotional intelligence. Just like people, if dogs can be happy, then surely they can be sad and lonely.

Throughout the Dog Project, I had been struck by how perfectly dogs and humans complemented each other. Humans, even with our powerful brains and capacity for abstract thought, are still slaves to our emotions, which dogs will pick up on and resonate with. And the most powerful emotion of all is love. Despite the complexities of human relationships, the fundamental attribute of love is empathy. To love, and be loved, is to feel what another feels and have that returned. It really is that simple. If people do this with each other, it seems perfectly natural for us to do it with animals. People become intensely attached to their pets. Every day, on my way to work, I pass a professional building with a sign advertising grief counseling for pet loss. It is not an exaggeration to say that for many people, their pets are their primary relationships and that they love their cats and dogs more than people. This is why it hurts so much when we lose them.

We were not a one-dog household. The grief at Lyra's passing had been profound, but the emptiness had been worse. Eventually, the entire family, including Callie, once again took a trip to the animal shelter. This time, Callie was there to help pick a new dog to join the family.

Walking down row after row of barking dogs, Callie look-alikes were everywhere. It seemed that every cage held a gaunt village dog. Black fur, white chest, tail in a C. And every single one had the breed listed as pit bull terrier mix. A feist by any other name. The temptation was strong to get a twin for Callie, but Helen insisted on a puppy. Something soft and cuddly — like Lyra, but different.

We zeroed in on a fluffy brindle puppy. He had a long snout and droopy lips and floppy ears that were too big for his head. No doubt about this one. He was a hound. Unlike his neighbors, he wasn't barking. I crumpled up a piece of paper and tossed it in the corner of his pen. He bounded over to it and brought it back to me. This was supposedly the single best test of puppy temperament. A puppy that retrieved an object indicated a predisposition to work with humans. I was sold.

Callie gave him a good sniffing and wagged her tail. It was unanimous.

Continuing our tradition of literary names, Helen and Maddy called him Cato, after the character from Suzanne Collins's *The Hunger Games*. Never mind that Cato was the most dangerous enemy to

Katniss Everdeen, the heroine of the novel. At least he was bold and single-minded of purpose.

Our Cato, though, was a goofball. Gangly and awkward, he ran around the house, tripping over his feet and doing somersaults. Because of his penchant for putting everything in his mouth, he was dubbed the "fur ball with teeth."

By the time Cato was six months old, his personality had begun to emerge. He seemed to move through the teething stage without too much destruction, although he had an obsession with the tags on clothing. He also liked to unravel toilet paper rolls and drag a trail of paper out of the bathroom.

Kat noted the eerie similarity to Newton.

"I think Cato is Newton reincarnated," she said. "Those are the exact same things Newton used to do." She was right. Even though Cato had been, in some way, a replacement for Lyra, he was closer to being a new Newton.

Helen, now thirteen years old, wanted to be primarily responsible for raising Cato.

"Do you know what that means?" I asked.

"I will have to let him out at night until he is housebroken."

"Yes."

"And I will have to train Cato to sit and stay and walk." Cato heard his name and jumped into Helen's lap. He started licking her face.

"You will feed him?" I asked.

"Uh-huh."

"And you will pick up his poops on walks?"

Helen hesitated and thought about it. "Umm, I don't know about that."

I signed up Helen and Cato for puppy class at CPT. Mark's class was a gentle introduction to basic training for both puppies and owners and let the puppies socialize with other dogs in a safe environment.

Helen beamed in delight when she learned how to get Cato to sit and lie down. Like Callie, his love of hot dogs made training a breeze.

Of course, there was more to raising a puppy than basic training. If I had learned nothing else in the Dog Project, it was how to communicate better. Dogs come ready-made to soak up the social rules of the household. It was our human inconsistencies that made it difficult for them.

Humans emit a constant stream of signals. We talk constantly. Our bodies are in motion. We wave our hands in wild patterns in feeble attempts to communicate emotions. It isn't at all clear how much of this verbal and physical chatter is actually necessary. I realized that Callie ignored most of the family's gesticulations, instead reserving her attention to the signals that carried useful information. I respected her regal demeanor. What, two years ago, I had mistaken for aloofness, I now understood to be an economy of attention. She had revealed herself to be capable of great feats of mentalizing when working with me as part of the Dog Project. If she wasn't interested in what I was saying, I realized that it was because I wasn't being clear in what I wanted.

After spending hours staring eyeball-to-eyeball with Callie, we had achieved a level of communication that I don't think I had ever had with a dog. Not even Newton. I had learned to read some of Callie's body language, especially her eyes. Her flicks of attention telegraphed what caught her interest. The photographs and video footage from the Dog Project made it obvious that the dogs' attention was focused on the humans. I hadn't noticed it at the time, but when I replayed the footage, it was impossible to ignore. The dogs were watching us, trying to figure out what we were thinking and how to shape their own behavior to fit in.

Consistency and clarity. That was the ticket. I resolved to be more consistent — with both dogs and humans alike.

After one of the puppy classes, Helen asked me, "Could Cato be in the Dog Project?"

"He's too young," I replied.

"How old does he have to be?"

"At least a year old."

"But," Helen opined, "he's really smart. I bet he could hold his head still."

"He probably could. But puppies' brains aren't fully grown. We wouldn't know how to compare his brain to an adult brain like Callie's."

Helen took this in. "Could I start training him so that he'll be ready by the time he's one year old?"

"Sure," I said. "But why do you want him to be in the Dog Project?"

Helen stroked Cato's head. "So I can know what he's thinking."

I smiled. I knew exactly how she felt.

Epilogue

Two years and two dogs. Two dogs scanned and two dogs gone. The Dog Project started as an idea born from the grief of losing our pug Newton but blossomed into something bigger than any of us could have expected. Up until that point, I had kept my feelings toward dogs mostly to myself. But after we published the initial results with Callie and McKenzie, there was an outpouring of support from people all over the world. I was moved by how strongly people wanted to know what their dogs were thinking.

One of the first people I heard from was Jessie Lendennie, a poet and managing editor at the publisher Salmon Poetry in Ireland. Jessie was kind enough to send me an anthology of poems, *Dogs Singing*, that she had compiled from poets all over the world. It is a remarkable tribute to the powerful effect that dogs have on people. I found it inspiring as the Dog Project progressed beyond just two dogs.

Moving forward was a risky move. We still had no funding to speak of. The only reason we got as far as we did was through the volunteer efforts of Mark and Melissa and all the people in my lab, especially Andrew. Scanner time still cost $500 an hour, and there were no freebies in that department. I had paid for scan costs out of discretionary research funds that I had accumulated over the years, but at the end of the hot dog and smell experiments, we had to ask ourselves: What now?

We had the only dogs in the world that were trained to go into an MRI. We could keep dreaming up questions to ask about how the canine brain worked, but there were limits in what we could learn from just two subjects. If the Dog Project was to continue to decipher what our furry friends think about us, the path was clear: we needed more dogs. If we had more dogs, we could sort out the questions about how many of the differences between Callie and McKenzie were because of their genetics, their environment, or just random day-to-day fluctuations in their mood, which surely must happen, just like humans. We all wanted to know about the differences in breeds.

Even though I wasn't sure how we would pay for all this, there was never any real question that I was going all in. I had always followed my passion and hunches in science rather than trying to fit my research program into whatever the funding agencies' hot topic of the year was. I didn't hesitate to commit all of my available resources to expand the Dog Project. I had faith that eventually people would soon realize that this wasn't a frivolous endeavor, and that deciphering what goes on in dogs' minds would tell us something about where humans came from and how we can live more harmoniously with these wonderful creatures.

The first order of business was to recruit the A-Team. Mark sent out an e-mail to everyone who had come through CPT in the past ten years. He put in calls to local veterinarians. We set a high bar. Callie and McKenzie had shown us what kind of dog could do this. Dogs had to be calm, good in novel environments, good with strangers, good with other dogs, inquisitive, unafraid of loud noises, able to wear earmuffs, and, above all, have a drive to learn new things.

We held tryouts. We tested the dog-human teams for their ability to learn new tasks, like going in the head coil and wearing earmuffs. We played recordings of the scanner noise, watching for any signs of anxiety. After hours of testing, we were still left with five new dogs and owners who were ready to commit to the project. Just as exciting,

Kady wearing earmuffs (above). Tigger in the head coil (below).
(*Helen Berns*)

the dogs represented a cross section of breeds. We had Kady, a Lab-golden mix who had washed out of therapy training for being too sensitive. There was Rocky, the miniature poodle; Caylin, another border collie; and Huxley, a Brittany mix. And finally, rounding out the motley crew was Tigger, a funny Boston terrier who reminded me a lot of Newton.

Mark honed our training plan, and we began holding weekly

Callie testing a neck coil.
(Helen Berns)

classes at CPT where we gradually acclimated the dogs to the MRI environment. In just a few months we had gone from two dogs to eight, and we were well on our way to boldly going where no dogs had gone before!

Callie continued as top dog. Whenever we added something new — a new experiment or a piece of equipment — Callie was the first to try it out. With her help, we discovered that we could obtain stronger signals from a dog's brain by using a coil designed for the human neck. With this type of coil, the pickup element was closer to the brain than in the birdcage.

In talking about the Dog Project, I have learned that people react in one of two possible ways. Dog people do not need any further explanation. They understand the desire to know what their dogs are thinking, especially how they love them. If anything, these folks wonder why nobody has done this before. The other type of person, possibly a dog owner but not actually a dog person, views this as a colossal waste of money. Shouldn't we be using these expensive MRI machines to improve human health?

It is a valid question, and the best way I can answer it is to say that through the Dog Project, we *are* improving the human condition. Although I had worked in neuroscience for almost twenty years, and the majority of my research had been funded by the National Institutes of Health to understand how the human reward system goes awry in addiction, more people have been positively impacted by our one experiment on two dogs' brains than the thousands of MRIs we had previously done in humans. Not everyone loves dogs, but for those who do — and that is about half the people in the United States — their dog's welfare is intimately tied to their own. If we can understand just a little bit of what goes on behind those puppy eyes, dog-human bonds can only become stronger.

There are already well-documented beneficial effects of living with animals. The Centers for Disease Control and Prevention notes that living with pets can decrease blood pressure, cholesterol, and triglyceride levels, as well as alleviate feelings of loneliness. Dogs, especially, provide opportunities for exercise and socialization.

As we move forward with the Dog Project, it is one of my dreams to really figure out what makes for a strong dog-human bond, what Konrad Lorenz called a "resonant dog" — a dog and a human who are fully in sync with each other. Using the reactivity of specific parts of a dog's brain to a human, we could gauge the strength of this bond and figure out activities to strengthen it, or better match people with dogs. Therapy animals could benefit as well, both in finding and training dogs to be the most effective at this important activity.

And while it is easy to see how we could use this information to improve human health, I think it is just as important to use this technology to improve the welfare of dogs. Although they are considered by many to be man's best friend, they are also still misunderstood, which I think is a result of many people's impression that dogs are barely domesticated wolves. I was disappointed when I encountered this attitude from an NIH official to whom I was proposing the expansion of the Dog Project to better understand how dogs decrease stress in humans. Instead of seeing how the dog's reward system is tied into the human's well-being, his response was "I imagine [the reward system] is maximally active when the dog is tearing *into* a human." I can only speculate that he had a bad childhood experience with dogs or that he had been reading too many werewolf stories.

The point is that we can use brain-imaging technology for our own benefit, but we can also train it on dogs for their benefit. We are only scratching the surface of figuring out what dogs know and what they feel. But we already know that the major cause of behavioral problems in modern dogs is separation anxiety. Dogs get attached to their humans, and, understandably, they get lonely when the people are gone. When they act out and destroy things, everyone suffers, but it is the dog that may end up at the shelter.

It may seem far-fetched that scanning dogs' brains could solve problems like this. But since a dog cannot tell us what is bothering him, peering into his mind may tell us what aspect of being separated from his human causes the most distress. Is it a matter of time or distance? How effective could webcams be in checking in with our dogs during the day? Currently, nobody knows how to best tap into dogs' perceptual systems through technology. Brain imaging could lead the way.

Beyond all the promise of new discovery, the aspect of the Dog Project of which I am most proud is how we treated the dogs. Of course, Callie and McKenzie were family members, but we treated them like humans in the hope that others who followed in their paw-

steps would be afforded the same respect and rights of self-determination. Until proven otherwise, I believe the right course of action
is to assume that dogs (and probably many other kinds of animals)
have a level of self-awareness and emotion that bears more in common with humans than we had ever anticipated.

Dogs are surely our first friends for always.

Notes

page

2. WHAT IT'S LIKE TO BE A DOG

16 What is it like to be a dog . . . : Thomas Nagel. "What is it like to be a bat?" *Philosophical Review* 83, no. 4 (October 1974): 435–450.
Many authors have written about the dog mind . . . : For a particularly good review see: John Bradshaw. *Dog Sense: How the New Science of Dog Behavior Can Make You a Better Friend to Your Pet* (New York: Basic Books, 2011).

17 Lupomorphism . . . : Adam Miklosi. *Dog Behaviour, Evolution, and Cognition* (Oxford and New York: Oxford University Press, 2007), p. 15.

18 Visual part of the brain and imagination . . . : Xu Cui et al. "Vividness of mental imagery: individual variability can be measured objectively." *Vision Research* 47, no. 4 (February 2007): 474–478.

4. PUPPY STEPS

34 Classical conditioning . . . : Steven R. Lindsay. *Handbook of Applied Dog Behavior and Training.* Vol. 1, *Adaptation and Learning* (Ames: Iowa State University Press, 2000).

6. RESONANT DOGS

52 Florence Nightingale . . . : Florence Nightingale. *Notes on Nursing: What It Is, and What It Is Not* (New York: D. Appleton, 1860), p. 103.

53 Demonstrating that dogs and animals in general can improve human health . . . : Lori S. Palley, P. Pearl O'Rourke, and Steven M. Niemi. "Mainstreaming animal-assisted therapy." *ILAR Journal* 51, no. 3 (2010): 199–207.
Children and pet therapy . . . : Kathie M. Cole et al. "Animal-assisted therapy in patients hospitalized with heart failure." *American Journal of Critical Care* 16, no. 6 (November 2007): 575–585. Elaine E. Lust et al. "Measuring clinical outcomes of animal-assisted therapy: impact on resident medication usage." *Consultant Phar-*

macist 22, no. 7 (July 2007): 580–585. Carie Braun et al. "Animal-assisted therapy as a pain relief intervention for children." *Complementary Therapies in Clinical Practice* 15, no. 2 (May 2009): 105–109.

Animal-assisted therapy patterns . . . : This is called a *meta-analysis* and was reported in: Janelle Nimer and Brad Lundahl. "Animal-assisted therapy: a meta-analysis." *Anthrozoos* 20, no. 3 (September 2007): 225–238.

54 Konrad Lorenz . . . : Konrad Lorenz. *Man Meets Dog*. Translated by Marjorie Kerr Wilson (New York, Tokyo, and London: Kodansha International, 1994).

Animals demonstrate an understanding of fairness . . . : Frans de Waal. *Our Inner Ape: A Leading Primatologist Explains Why We Are Who We Are* (New York: Riverhead, 2005).

Resonance dog . . . : Lorenz, *Man Meets Dog*, p. 76.

7. LAWYERS GET INVOLVED

66 Rabies in the United States . . . : "Human Rabies." Centers for Disease Control and Prevention, last modified May 3, 2012. http://www.cdc.gov/rabies/location/usa/surveillance/human_rabies.html.

8. THE SIMULATOR

69 First investigation of dogs' hearing . . . : E. A. Lipman and J. R. Grassi. "Comparative auditory sensitivity of man and dog." *American Journal of Psychology* 55, no. 1 (January 1941): 84–89.

9. BASIC TRAINING

80 Puppies and social learning . . . : Leonore L. Adler and Helmut E. Adler. "Ontogeny of observational learning in the dog (*Canis familiaris*)." *Developmental Psychobiology* 10, no. 3 (May 1977): 267–271. Cf. A. Miklosi, *Dog Behaviour*.

Puppies that watched their mother . . . : J. M. Slabbert and O. Anne E. Rasa. "Observational learning of an acquired maternal behaviour pattern by working dog pups: an alternative training method?" *Applied Animal Behaviour Science* 53, no. 4 (July 1997): 309–316.

85 Mutt Muffs . . . : Mutt Muffs, accessed December 20, 2012. http://www.safeandsoundpets.com/index.html.

11. THE CARROT OR THE STICK?

100 Cesar Millan and pack leader . . . : Cesar Millan and Melissa Jo Peltier. *Be the Pack Leader: Use Cesar's Way to Transform Your Dog . . . and Your Life* (New York: Harmony Books, 2007).

12. DOGS AT WORK

107 Dogs in the workplace . . . : Randolph T. Barker et al. "Preliminary investigation of employee's dog presence on stress and organizational perceptions." *International Journal of Workplace Health Management* 5, no. 1 (2012): 15–30.

108 Chronically high levels of cortisol . . . : Robert M. Sapolsky. *Why Zebras Don't Get Ulcers*, 3rd ed. (New York: Henry Holt, 1994).

Google's dog policy . . . : "Code of Conduct." Google Investor Relations, last modified April 24, 2012. http://investor.google.com/corporate/code-of-conduct .html#toc-dogs.

109 Dog-friendly businesses . . . : The website dogfriendly.com has a user-contributed list of companies that allow dogs.

Charles Darwin . . . : Charles Darwin. *The Expression of the Emotions in Man and Animals*. Introduction, afterword, and commentaries by Paul Ekman. 4th ed. (Oxford: Oxford University Press, 2009), pp. 55–56.

111 Darwin's work was forgotten for more than a century . . . : The situation has begun to change, in large part because of the efforts of Paul Ekman, a psychologist who has extensively studied the facial expressions in humans, and Frans de Waal, an ethologist who studies primate behavior.

There have been a few exceptions . . . : Marc Bekoff, an ethologist at the University of Colorado at Boulder, has spent much of his career extending Darwin's work. Bekoff has argued strenuously for the recognition of animal emotions: Marc Bekoff. *The Emotional Lives of Animals: A Leading Scientist Explores Animal Joy, Sorrow, and Empathy — and Why They Matter* (Novato, CA: New World Library, 2007).

Jaak Panksepp . . . : Jaak Panksepp. *Affective Neuroscience: The Foundations of Human and Animal Emotions* (New York: Oxford University Press, 1998).

112 Breaking emotion down to fundamental components . . . : Stanley Schachter and Jerome E. Singer. "Cognitive, social, and physiological determinants of emotional state." *Psychological Review* 69, no. 5 (September 1962): 379–399.

Circumplex model . . . : James A. Russell. "A circumplex model of affect." *Journal of Personality and Social Psychology* 39, no. 6 (1980): 1161–1178.

113 The "seeking" system . . . : Jaak Panksepp. "The basic emotional circuits of mammalian brains: do animals have affective lives?" *Neuroscience and Biobehavioral Reviews* 35, no. 9 (October 2011): 1791–1804.

14. BIG QUESTIONS

126 Electrical stimulation of dog brains . . . : Gustav Fritsch and Eduard Hitzig. "Ueber die elektrische Erregbarkeit des Grosshirns" [Electric excitability of the cerebrum]. *Archiv fuer Anatomie, Physiologie und Wissenschaftliche Medicin* 37

(1870): 300–322. T. Gorska. "Functional organization of cortical motor areas in adult dogs and puppies." *Acta Neurobiologiae Experimentalis* 34, no. 1 (1974): 171–203.

127 Caudate nucleus and reward . . . : Reward processing is most closely associated with the nucleus accumbens, which is a subregion of the caudate. This region is also called the *ventral striatum*. For brevity, I refer to both as the caudate.
Wolfram Schultz and measurement of caudate activity . . . : Wolfram Schultz et al. "Neuronal activity in the monkey ventral striatum related to the expectation of reward." *Journal of Neuroscience* 12, no. 12 (December 1992): 4595–4610.

16. A NEW WORLD

155 Dog brain images from University of Minnesota Canine Brain MRI Atlas (http://vanat.cvm.umn.edu/mriBrainAtlas/) by T. F. Fletcher and T. C. Saveraid, 2009.

156 Reverse inference . . . : Russell A. Poldrack. "The role of fMRI in cognitive neuroscience: where do we stand?" *Current Opinion in Neurobiology* 18, no. 2 (April 2008): 223–227.
Reverse inference in the caudate . . . : Dan Ariely and Gregory S. Berns. "Neuromarketing: the hope and hype of neuroimaging in business." *Nature Reviews Neuroscience* 11, no. 4 (April 2010): 284–292.

157 Love and the caudate . . . : Arthur Aron et al. "Reward, motivation, and emotion systems associated with early-stage intense romantic love." *Journal of Neurophysiology* 94, no. 1 (July 2005): 327–337.

17. PEAS AND HOT DOGS

164 Side preference in dogs . . . : Camille Ward and Barbara B. Smuts. "Quantity-based judgments in the domestic dog (*Canis lupus familiaris*)." *Animal Cognition* 10, no. 1 (January 2007): 71–80.

18. THROUGH A DOG'S EYES

169 Signal-to-noise ratio . . . : The SNR increases roughly by a factor of $\sqrt{N}$, where N is the number of repetitions. For example, 100 repetitions would increase the SNR by a factor of 10.

173 Dogs used attentional cues from humans . . . : Márta Gácsi et al. "Are readers of our face readers of our minds? Dogs (*Canis familiaris*) show situation-dependent recognition of human's attention." *Animal Cognition* 7, no. 3 (July 2004): 144–153.
Dogs are sensitive to the social context . . . : Juliane Kaminski et al. "Domestic dogs are sensitive to a human's perspective." *Behaviour* 146, no. 7 (2009): 979–998.
Alexandra Horowitz. "Theory of mind in dogs? Examining method and concept." *Learning and Behavior* 39, no. 4 (December 2011): 314–317.

174 Knowing how to read people and how to behave in different social settings is the

difference between success and failure . . . : Gregory Berns. *Iconoclast: A Neuroscientist Reveals How to Think Differently* (Boston: Harvard Business School Press, 2008).

19. EUREKA!

182 Nothing in the brain implies an understanding of meaning . . . : When I first presented the findings to a group of psychologists, this is exactly what they said.

Dogs' ability to intuit the meaning of human social signals . . . : Brian Hare and Michael Tomasello. "Human-like social skills in dogs?" *Trends in Cognitive Sciences* 9, no. 9 (September 2005): 439–444. Brian Hare, Josep Call, and Michael Tomasello. "Communication of food location between human and dog (*Canis familiaris*)." *Evolution of Communication* 2, no. 1 (1998): 137–159. See also A. Miklósi et al. "Use of experimenter-given cues in dogs." *Animal Cognition* 1, no. 2 (1998): 113–121.

Social cognition of wolves and chimpanzees . . . : Brian Hare et al. "The domestication of social cognition in dogs." *Science* 298, no. 5598 (November 2002): 1634–1636. See also Brian Hare and Vanessa Woods. *The Genius of Dogs: How Dogs Are Smarter than You Think* (New York: Dutton, 2013).

183 Kool-Aid experiment . . . : Giuseppe Pagnoni et al. "Activity in human ventral striatum locked to errors of reward prediction." *Nature Neuroscience* 5, no. 2 (2002): 97–98.

Dysfunctional caudate in addiction . . . : Nora D. Volkow et al. "Addiction: beyond dopamine reward circuitry." *Proceedings of the National Academy of Sciences of the United States of America* 108, no. 37 (September 2011): 15037–15042.

Bonus effect in the caudate to social cues . . . : James K. Rilling et al. "A neural basis for social cooperation." *Neuron* 35, no. 2 (July 18, 2002): 395–405. I. Aharon et al. "Beautiful faces have variable reward value: fMRI and behavioral evidence." *Neuron* 32, no. 3 (November 8, 2001): 537–551.

20. DOES MY DOG LOVE ME?

186 We had finished the first scientific paper . . . : Gregory S. Berns, Andrew M. Brooks, and Mark Spivak. "Functional MRI in Awake Unrestrained Dogs." *Public Library of Science ONE* 7, no. 5 (2012): e38027.

190 Mirror neurons . . . : Giacomo Rizzolatti and Luigi Craighero. "The mirror-neuron system." *Annual Review of Neuroscience* 27 (2004): 169–192.

191 Mirror neurons are the basis of empathy . . . : Marco Iacoboni and Mirella Dapretto. "The mirror neuron system and the consequences of its dysfunction." *Nature Reviews Neuroscience* 7 (December 2006): 942–951.

Imitation and empathy . . . : Marco Iacoboni. "Imitation, empathy, and mirror neurons." *Annual Review of Psychology* 60 (January 2009): 653–670.

192 Brain activation to silent movie of dogs barking . . . : Kaspar Meyer et al. "Predicting visual stimuli on the basis of activity in auditory cortices." *Nature Neuroscience* 13, no. 6 (June 2010): 667–668.

21. WHAT'S THAT SMELL?

196 Dog's sense of smell is 100,000 times as sensitive . . . : John Bradshaw. *Dog Sense*.

197 Smell and control of movement . . . : Joel D. Mainland et al. "Olfactory impairments in patients with unilateral cerebellar lesions are selective to inputs from the contralesional nostril." *Journal of Neuroscience* 25, no. 27 (July 6, 2005): 6362–6371.

198 Dogs can differentiate their own urine . . . : Marc Bekoff. "Observations of scent-marking and discriminating self from others by a domestic dog (*Canis familiaris*): tales of displaced yellow snow." *Behavioural Processes* 55, no. 2 (August 15, 2001): 75–79.

202 McKenzie was not having a good day . . . : McKenzie came back three weeks later, after Mark and Melissa had practiced with her. She then performed like a champ and sat for more than seven hundred scans.

204 Dog brain activation looked like human activation to people they love . . . : Aron et al. "Reward, motivation, and emotion systems."

22. FIRST FRIEND

207 Most of the dogs in the world are village dogs . . . : See the classic book on this topic: Raymond Coppinger and Lorna Coppinger. *Dogs: A New Understanding of Canine Origin, Behavior, and Evolution* (Chicago: University of Chicago Press, 2001).

208 Rapidity with which the dog has changed form . . . : Coppinger and Coppinger, *Dogs*, p. 297.

211 Dogs take the form of the people they live with . . . : Lance Workman, a psychologist at Bath Spa University in Britain, has studied both the physical resemblance of dogs to their owners as well as their personalities and finds evidence of such a relationship.

212 "Less-is-more" approach to dog communication . . . : Patricia B. McConnell. *The Other End of the Leash: Why We Do What We Do Around Dogs* (New York: Ballantine Books, 2002).

213 Supreme Court and neuroscience . . . : *Graham v. Florida*, 560 US (2010).

24. WHAT DOGS ARE REALLY THINKING

226 Self-domestication . . . : Hare and Woods. *Genius of Dogs*.

228 Wolves require a prodigious amount of food . . . : John Bradshaw. *Dog Sense*. Prehistoric cave art . . . : Pat Shipman. *The Animal Connection: A New Perspective on What Makes Us Human* (New York: W. W. Norton, 2011), p. 227.

Acknowledgments

First, and foremost, I owe special gratitude to Andrew Brooks and Mark Spivak. Andrew took a risk by working on the project, giving up time from his PhD studies to follow our dream of reading the mind of a dog. Without his contributions and tireless work ethic, none of this could have happened. And without Mark, we would never have been able to train the dogs to go into the scanner. Mark took on the project because it sounded fun and interesting but ended up volunteering countless hours honing our training protocols. More than a dog trainer, Mark made invaluable contributions to the science as we plunged forward. I couldn't ask for better colleagues than Andrew and Mark.

The other people in the lab contributed in so many ways, large and small. Someday I will look back upon these two years and realize that it was a golden time, made special by the luck of having the right people at just the right time. Thanks to: Jan Barton, Kristina Blaine, Monica Capra, Gavin Ekins, David Freydkin, Lisa LaViers, Melanie Pincus, Michael Prietula, and Brandon Pye.

Outside of the lab, I am grateful to Larry Iten, director of the Emory IACUC, for not hanging up on me when I called him to propose the Dog Project. Larry helped me shepherd the project through the labyrinth of animal research regulations. Sarah Putney, director of the IRB, helped draft the consent form for dog owners and talked us through the implications of treating dogs like children in research. The veterinary staff at Emory has been great, with special thanks to Deborah Mook, Michael Huerkamp, and especially Rebeccah Hunter, who figured out how to keep the earmuffs on the dogs. At the scanner, I am grateful for the MR wizardry of Robert Smith, Lei Zhou, and Sinyeob Ahn, who were crucial in figuring out how to program the MRI to scan dogs.

I am eternally grateful to everyone who volunteered their dogs and their time to be part of the Dog Project. Melissa Cate was the first to join the team with her dog McKenzie. Without them, this would have been no more than a cute dog trick for Callie. Thanks also to the members of the A-Team: Patricia King and Kady, Lorrie Backer and Caylin, Aliza Levenson and Tigger, Melanie Pincus and Huxley, and Richard Fischhof and Rocky.

Thanks to Jim Levine for encouraging me to chronicle the experiences of the Dog Project, helping develop the idea into a book, and matching me with David Moldawer at Amazon, who helped take the book to the finish line. Thanks to Bryan Meltz for her amazing photographic skill.

And finally, thanks to Kat, Helen, and Maddy for living with the Dog Project. I promise to remove the simulator from the living room someday.

GREGORY BERNS, M.D., Ph.D., is the Distinguished Professor of Neuroeconomics at Emory University. His research has been featured in the *New York Times*, the *Wall Street Journal*, *Forbes*, the *Los Angeles Times*, *Nature*, *Money*, *New Scientist*, *Psychology Today*, and on CNN, NPR, ABC, and the BBC. He lives in Atlanta, Georgia, with his wife, two children, and three dogs.

 @gberns